Part 3:   Fluctuations Option C
           (Formula adjustment)

Articles of Agreement

This Agreement is made the _____ 20 _____

Between **The Employer** _____

_____

_____ (Company No. _____)[1]

of/whose registered office is at _____

_____

_____

And **The Contractor** _____

_____

_____ (Company No. _____)[1]

of/whose registered office is at _____

_____

_____

[1] Where the Employer or Contractor is neither a company incorporated under the Companies Acts nor a company registered under the laws of another country, delete the references to Company number and registered office. In the case of a company incorporated outside England and Wales, particulars of its place of incorporation should be inserted immediately before its Company number.
As to execution by foreign companies and matters of jurisdiction, see the Guide.

Recitals

Whereas

First   the Employer wishes to have the following work carried out[2]:

_____

_____

at _____

_____ ('the Works')
and has had drawings and bills of quantities prepared which show and describe the work to be done;

Second   the Contractor has supplied the Employer with a fully priced copy of the bills of quantities, which for identification has been signed or initialled by or on behalf of each Party ('the Contract Bills');

and has provided the Employer with the priced schedule of activities annexed to this Contract ('the Activity Schedule')[3];

Third   the drawings are numbered/listed in _____

_____ annexed to this Contract ('the Contract Drawings')
and have for identification been signed or initialled by or on behalf of each Party[4];

Fourth   for the purposes of the Construction Industry Scheme (CIS) under the Finance Act 2004, the status of the Employer is, as at the Base Date, that stated in the Contract Particulars;

Fifth   the Employer has provided the Contractor with a schedule ('the Information Release Schedule') which states the information the Architect/Contract Administrator will release and the time of that release[5];

Sixth   the division of the Works into Sections is shown in the Contract Bills and/or the Contract Drawings or in such other documents as are identified in the Contract Particulars[6];

---

[2]   State nature and location of intended works.
[3]   Delete these lines if a priced Activity Schedule is not provided.
     In the Activity Schedule, each activity should be priced, so that the sum of those prices equals the Contract Sum excluding Provisional Sums and the value of work for which Approximate Quantities are included in the Contract Bills.
[4]   State the identifying numbers of the Contract Drawings or identify the schedule of drawings or other document listing them, which should be annexed to this Contract, and make the appropriate deletions. The drawings themselves should be signed or initialled by or on behalf of each Party.
[5]   Delete the Fifth Recital if an Information Release Schedule is not provided.
[6]   Delete the Sixth Recital if the Works are not divided into Sections.

The Seventh to Tenth Recitals apply only where there is a Contractor's Designed Portion

Seventh   the Works include the design and construction of[7] _____

_____

_____('the Contractor's Designed Portion');

Eighth   the Employer has supplied to the Contractor documents showing and describing or otherwise stating his requirements for the design and construction of the Contractor's Designed Portion ('the Employer's Requirements');

Ninth   in response to the Employer's Requirements the Contractor has supplied to the Employer:

- documents showing and describing the Contractor's proposals for the design and construction of the Contractor's Designed Portion ('the Contractor's Proposals'); and

- an analysis of the portion of the Contract Sum relating to the Contractor's Designed Portion ('the CDP Analysis');

Tenth   the Employer has examined the Contractor's Proposals and, subject to the Conditions, is satisfied that they appear to meet the Employer's Requirements. The Employer's Requirements, the Contractor's Proposals and the CDP Analysis have each for identification been signed or initialled by or on behalf of each Party and particulars of each are given in the Contract Particulars;

---

[7]   State nature of work in the Contractor's Designed Portion, or delete these four Recitals if not applicable. If the space here is insufficient a separate list should be prepared, signed or initialled by or on behalf of each Party and identified here, either as a specified Annex to this Contract or by its reference number, date or other identifier.

Now it is hereby agreed as follows

Article 1: Contractor's obligations

The Contractor shall carry out and complete the Works in accordance with the Contract Documents.

Article 2: Contract Sum

The Employer shall pay the Contractor at the times and in the manner specified in the Conditions the VAT-exclusive sum of

_____

_____ (£ _____._____) ('the Contract Sum')

or such other sum as shall become payable under this Contract.

Article 3: Architect/Contract Administrator

For the purposes of this Contract the Architect/Contract Administrator is

_____

of _____

_____

or, if he ceases to be the Architect/Contract Administrator, such other person as the Employer shall nominate in accordance with clause 3·5 of the Conditions.

Article 4: Quantity Surveyor

For the purposes of this Contract the Quantity Surveyor is

_____

of _____

_____

or, if he ceases to be the Quantity Surveyor, such other person as the Employer shall nominate in accordance with clause 3·5 of the Conditions.

Article 5: CDM Co-ordinator

The CDM Co-ordinator for the purposes of the CDM Regulations is the Architect/Contract Administrator

(or)[8] _____

of _____

_____

or, if he ceases to be the CDM Co-ordinator, such other person as the Employer shall appoint pursuant to regulation 14(1) of those regulations.

Article 6: Principal Contractor

The Principal Contractor for the purposes of the CDM Regulations is the Contractor

(or)[8] _____

of _____

_____

or, if he ceases to be the Principal Contractor, such other contractor as the Employer shall appoint pursuant to regulation 14(2) of those regulations.

Article 7: Adjudication

If any dispute or difference arises under this Contract, either Party may refer it to adjudication in accordance with clause 9·2.[9]

Article 8: Arbitration

Where Article 8 applies[10], then, subject to Article 7 and the exceptions set out below, any dispute or difference between the Parties of any kind whatsoever arising out of or in connection with this Contract, whether before, during the progress or after the completion or abandonment of the Works or after the termination of the Contractor's employment, shall be referred to arbitration in accordance with clauses 9·3 to 9·8 and the JCT 2005 edition of the Construction Industry Model Arbitration Rules (CIMAR). The exceptions to this Article 8 are:

- any disputes or differences arising under or in respect of the Construction Industry Scheme or VAT, to the extent that legislation provides another method of resolving such disputes or differences; and

- any disputes or differences in connection with the enforcement of any decision of an Adjudicator.

Article 9: Legal proceedings[10]

Subject to Article 7 and (where it applies) to Article 8, the English courts shall have jurisdiction over any dispute or difference between the Parties which arises out of or in connection with this Contract.

---

[8]   Insert the name of the CDM Co-ordinator only where the Architect/Contract Administrator is not to fulfil that role, and that of the Principal Contractor only if that is to be a person other than the Contractor. If the project is not notifiable under the CDM Regulations 2007 (i.e. a project which is not likely to involve more than 30 days, or 500 person days, of construction work or which is being carried out for a homeowner as a purely domestic project), delete Articles 5 and 6 in their entirety.

[9]   As to adjudication in cases where the Employer is a residential occupier within the meaning of section 106 of the Housing Grants, Construction and Regeneration Act 1996, see the Guide.

[10]   If it is intended, subject to the right of adjudication and exceptions stated in Article 8, that disputes or differences should be determined by arbitration and not by legal proceedings, the Contract Particulars **must** state that Article 8 and clauses 9·3 to 9·8 apply and the words "do not apply" **must** be deleted. If the Parties wish any dispute or difference to be determined by the courts of another jurisdiction the appropriate amendment should be made to Article 9 (see also clause 1·12 and Schedule 5 Parts 1 and 2).

Contract Particulars

Note: An asterisk * indicates text that is to be deleted as appropriate.

Part 1: General

Clause etc.	*Subject*	
Fourth Recital and clause 4·7	Construction Industry Scheme (CIS)	Employer at the Base Date * is a 'contractor'/is not a 'contractor' for the purposes of the CIS
Sixth Recital	Description of Sections (if any) *(If not shown or described in the Contract Drawings or Contract Bills, state the reference numbers and dates or other identifiers of documents in which they are shown.)*[11]	_____ _____ _____ _____
Eighth Recital	Employer's Requirements *(State reference numbers and dates or other identifiers of documents in which these are contained.)*[11]	_____ _____ _____ _____ _____
Ninth Recital	Contractor's Proposals *(State reference numbers and dates or other identifiers of documents in which these are contained.)*[11]	_____ _____ _____ _____ _____
Ninth Recital	CDP Analysis *(State reference numbers and dates or other identifiers of documents in which this is contained.)*[11]	_____ _____ _____ _____ _____
Article 8	Arbitration *(If neither entry is deleted, Article 8 and clauses 9·3 to 9·8 will not apply. If disputes and differences are to be determined by arbitration and not by legal proceedings, it <u>must</u> be stated that Article 8 and clauses 9·3 to 9·8 apply.)*[12]	Article 8 and clauses 9·3 to 9·8 (*Arbitration*) * apply/do not apply

[11] If the relevant document or set of documents takes the form of an Annex to this Contract, it is sufficient to refer to that Annex.
[12] On factors to be taken into account by the Parties in considering whether disputes are to be determined by arbitration or by legal proceedings, see the Guide. See also footnote [10].

| 1·1 | Base Date | _____ |

1·1	CDM Planning Period[13]	shall mean the period of
		_____ * days/weeks
		* ending on the Date of Possession/
		* beginning/ending on
		_____ 20 ___

| 1·1 | Date for Completion of the Works
(where completion by Sections does not apply) | _____ |

	Sections: Dates for Completion of Sections[14]	Section ____ : _____
		Section ____ : _____
		Section ____ : _____

| 1·7 | Addresses for service of notices etc. by the Parties
(If none is stated, the address in each case, unless and until otherwise agreed and subject to clause 1·7·2, shall be that shown at the commencement of the Agreement.)[15] | Employer _____ |
		_____
		_____
		(Fax Number)_____
		Contractor _____
		_____
		_____
		(Fax Number) _____

| 1·8 | Electronic communications
The communications that may be made electronically and the format in which those are to be made[14]
(If none are identified, all communications are to be in writing, unless subsequently agreed otherwise.) | * are as follows/ |
		* are set out in the following document
		_____
		_____
		_____
		_____

---

[13] Under the CDM Regulations 2007 every client is expressly required to allocate sufficient time (the CDM Planning Period) prior to the commencement of construction to enable contractors and others to carry out necessary CDM planning and preparation. There may be cases where that planning and preparation needs to be completed earlier than the Date of Possession and adaptation of the entries may be needed where there are Sections.

[14] Continue on further sheets if necessary, which should be signed or initialled by or on behalf of each Party and then be annexed to this Contract.

[15] As to service of notices etc. outside the United Kingdom, see the Guide.

| 2·4 | Date of Possession of the site
(where possession by Sections does not apply) | _____ 20____ |

	Sections: Dates of Possession of Sections[14]	Section ____ : _____ 20____
		Section ____ : _____ 20____
		Section ____ : _____ 20____

2·5	Deferment of possession of the site *(where possession by Sections does not apply)*	Clause 2·5
	*	applies/does not apply
		Maximum period of deferment (if less than 6 weeks) is
		_____

	Sections: deferment of possession of Sections	Clause 2·5
	*	applies/does not apply
		Maximum period of deferment (if less than 6 weeks) is[14]
		Section ____ : _____
		Section ____ : _____
		Section ____ : _____

| 2·19·3 | Contractor's Designed Portion: limit of Contractor's liability for loss of use etc. (if any) | £ _____ |

| 2·32·2 | Liquidated damages
(where completion by Sections does not apply) | at the rate of |
| | | £ _____ per _____ |

	Sections: rate of liquidated damages for each Section[14]	Section ____ : £ _____ per _____
		Section ____ : £ _____ per _____
		Section ____ : £ _____ per _____

2·37	Sections: Section Sums[14]	Section ____ : £ _____
		Section ____ : £ _____
		Section ____ : £ _____

| 2·38 | Rectification Period
(where completion by Sections does not apply)
(If no other period is stated, the period is 6 months.) | _____ months from the date of practical completion of the Works |

Sections: Rectification Periods[14]
(If no other period is stated, the period is 6 months.)

Section _____ : _____ months

Section _____ : _____ months

Section _____ : _____ months
from the date of practical completion of each Section

4·8 Advance payment
(Not applicable where the Employer is a Local Authority)

Clause 4·8
* applies/does not apply

If applicable:
the advance payment will be[16]

£ _____ /

_____ per cent of the Contract Sum

and will be paid to the Contractor on

_____ ;

it will be reimbursed to the Employer in the following amount(s) and at the following time(s)

_____

_____

_____

_____

4·8 Advance Payment Bond
(Not applicable where the Employer is a Local Authority)
(Where an advance payment is to be made, an advance payment bond is required unless stated that it is not required.)

An advance payment bond
* is/is not required

4·9·2 Dates of issue of Interim Certificates
(If none are stated, Interim Certificates are to be issued at intervals not exceeding one month up to the date of practical completion of the Works, or the date within one month thereafter; the first Interim Certificate is to be issued within one month of the Date of Possession.)

The first date is:

_____

and thereafter the same date in each month or the nearest Business Day in that month[17]

4·17·4 Listed Items – uniquely identified
(Delete the entry if no bond is required.)

* For uniquely identified Listed Items a bond as referred to in clause 4·17·4 in respect of payment for such items is required for

£ _____

[16] Insert either a monetary amount or a percentage figure, delete the alternative and complete the other required details.
[17] The first date should not be more than one month after the Date of Possession. Where it is intended that Interim Certificates be issued on the last day of each month, the entry may be completed/amended to read "the last day of *(insert month)* and thereafter the last day in each month or the nearest Business Day in that month."

| 4·17·5 | Listed Items – not uniquely identified
(Delete the entry if clause 4·17·5 does not apply.) | * | For Listed Items that are not uniquely identified a bond as referred to in clause 4·17·5 in respect of payment for such items is required for

£ _____ |

| 4·19 | Contractor's Retention Bond
(Not applicable where the Employer is a Local Authority)
(Not applicable unless stated to apply and relevant particulars are given below) | Clause 4·19
* applies/does not apply

If clause 4·19 applies, the maximum aggregate sum for the purposes of clause 2 of the bond is

£ _____

For the purposes of clause 6·3 of the bond, the expiry date shall be

_____ |

| 4·20·1 | Retention Percentage
(The percentage is 3 per cent unless a different rate is stated.) | _____ per cent |

4·21 and Schedule 7	Fluctuations Options[18] *(If no Fluctuations Option is selected, Option A applies.)*	Schedule 7: * Fluctuations Option A applies/ * Fluctuations Option B applies/ * Fluctuations Option C applies
	Percentage addition for Fluctuations Option A, paragraph A·12 or Option B, paragraph B·13	_____ per cent
	Formula Rules for Fluctuations Option C, paragraph C·1·2	Rule 3: Base Month _____ 20_____
	(For Local Authorities only)	Rule 3: Non-Adjustable Element _____ per cent
	(Unless Part II is stated to apply, Part I applies.)	Rules 10 and 30(i): * Part I/Part II of section 2 of the Formula Rules applies[19]

| 6·4·1·2 | Contractor's insurance – injury to persons or property
Insurance cover *(for any one occurrence or series of occurrences arising out of one event)* | £ _____ |

---

[18] Delete all but one.
[19] The Part to be deleted depends upon which method of formula adjustment (Part I – Work Category Method or Part II – Work Group Method) is applicable.

6·5·1	Insurance – liability of Employer *(Not required unless it is stated that it may be required and the minimum amount of indemnity is stated)*	Insurance * may be required/is not required Minimum amount of indemnity for any one occurrence or series of occurrences arising out of one event £ _____ [20]
6·7 and Schedule 3	Insurance of the Works – Insurance Options[18][21]	Schedule 3: * Insurance Option A applies/ * Insurance Option B applies/ * Insurance Option C applies
6·7 and Schedule 3 Insurance Option A (paragraphs A·1 and A·3), B (paragraph B·1) or C (paragraph C·2)	Percentage to cover professional fees *(If no other percentage is stated, it shall be 15 per cent.)*	_____ per cent
6·7 and Schedule 3 Insurance Option A (paragraph A·3)	Annual renewal date of insurance *(as supplied by the Contractor)*	_____
6·11	Contractor's Designed Portion (CDP) Professional Indemnity insurance Level of cover *(If an alternative is not selected the amount shall be the aggregate amount for any one period of insurance. A period of insurance for these purposes shall be one year unless otherwise stated.)* *(If no amount is stated, insurance under clause 6·11 shall not be required.)*	Amount of indemnity required * relates to claims or series of claims arising out of one event/ * is the aggregate amount for any one period of insurance and is £ _____
	Level of cover for pollution/contamination claims *(If none is stated, the required level of cover shall be the full amount of the indemnity cover stated above.)*	£ _____
	Expiry of required period of CDP Professional Indemnity insurance *(If no period is selected, the expiry date shall be 6 years from the date of practical completion of the Works.)*	* 6 years/ * 12 years/ * _____ years (not exceeding 12 years)

---

[20] If the indemnity is to be for an aggregate amount and not for any one occurrence or series of occurrences the entry should be amended to make this clear.

[21] Obtaining Terrorism Cover, which is necessary in order to comply with the requirements of Insurance Option A, B or C, will involve an additional premium and may in certain situations be difficult to effect. Where a difficulty arises discussion should take place between the Parties and their insurance advisers. See the Guide.

| 6·13 | Joint Fire Code | The Joint Fire Code |
| | | * applies/does not apply[22] |

| | If the Joint Fire Code applies, state whether the insurer under Schedule 3, Insurance Option A, B or C (paragraph C·2) has specified that the Works are a 'Large Project': | * Yes/No[22] |

| 6·16 | Joint Fire Code – amendments/revisions *(The cost shall be borne by the Contractor unless otherwise stated.)* | The cost, if any, of compliance with amendment(s) or revision(s) to the Joint Fire Code shall be borne by |
| | | * the Employer/the Contractor |

| 7·2 | Assignment/grant by Employer of rights under clause 7·2 *(If neither entry is deleted, clause 7·2 will apply.)* | Clause 7·2 |
| | | * applies/does not apply |

| | Sections: rights under clause 7·2 *(If clause 7·2 applies, amend the entry if rights under that clause are to apply to certain Sections only.)* | * Rights under clause 7·2 apply to each Section |

| 8·9·2 | Period of suspension *(If none is stated, the period is 2 months.)* | _____ |

| 8·11·1·1 to 8·11·1·5 | Period of suspension *(If none is stated, the period is 2 months.)* | _____ |

| 9·2·1 | Adjudication[23] | The Adjudicator is _____ |

| | Nominator of Adjudicator – where no Adjudicator is named or where the named Adjudicator is unwilling or unable to act (whenever that is established)[24]

(Where an Adjudicator is not named and a nominator has not been selected, the nominator shall be one of the nominators listed opposite selected by the Party requiring the reference to adjudication.) | President or a Vice-President or Chairman or a Vice-Chairman:
* Royal Institute of British Architects
* The Royal Institution of Chartered Surveyors
* Construction Confederation
* National Specialist Contractors Council
* Chartered Institute of Arbitrators |

---

[22] Where Insurance Option A applies these entries are made on information supplied by the Contractor.
[23] The Parties should either name the Adjudicator and select the nominator or, alternatively, select only the nominator.
The Adjudication Agreement (Adj) and the Adjudication Agreement (Named Adjudicator) (Adj/N) have been prepared by JCT for use when appointing an Adjudicator.
[24] Delete all but one of the nominating bodies asterisked.

9·4·1   Arbitration[25]

Appointor of Arbitrator (and of any replacement)[26]

(If no appointor is selected, the appointor shall be the President or a Vice-President of the Royal Institute of British Architects.)

President or a Vice-President:

*   Royal Institute of British Architects
*   The Royal Institution of Chartered Surveyors
*   Chartered Institute of Arbitrators

---

[25]   This only applies where the Contract Particulars state (against the reference to Article 8) that Article 8 and clauses 9·3 to 9·8 *(Arbitration)* apply.

[26]   Delete all but one of the bodies asterisked.

Part 2: Third Party Rights and Collateral Warranties

If such rights or warranties are required from the Contractor and the particulars required by Part 2 (A) to (D) are set out in a separate document, state here the reference number, date and/or other identifier of that document[27]

_____

_____

(or)

complete the particulars below:

P&T Rights Particulars

(A) Identity of Purchasers/Tenants on whom P&T Rights may be conferred, and whether (in the case of the Contractor) those rights are to be conferred as third party rights (clause 7A) or by Collateral Warranty (clause 7C)[28]

Clauses 7A, 7C and 7E of the Conditions	Name, class or description of person[29]	The part of the Works to be purchased or let	State in each case which of clause 7A or 7C is to apply
	_____	_____	_____

(Where no persons are identified by name, class or description either above or in a document identified above, P&T Rights cannot be conferred either under clause 7A (as Third Party Rights) or under clause 7C (by Collateral Warranty – CWa/P&T). If in relation to an identified person it is not stated whether clause 7A (Third Party Rights) or clause 7C (Collateral Warranty) applies, clause 7A shall apply.)

---

[27]   The document should for identification be signed or initialled by or on behalf of each Party and annexed to this Contract.
[28]   The Contractor may be required to grant rights either as Third Party Rights or Collateral Warranties. In the case of Sub-Contractors, provision is made only for the grant of Collateral Warranties – see the Guide.
[29]   As to the Contracts (Rights of Third Parties) Act 1999 and identification of beneficiaries, see the Guide.

Paragraph of Schedule 5, Part 1 or Clause of CWa/P&T[30]	**(B) P&T Rights from the Contractor**	
1·1·2	Applicability of paragraph/clause 1·1·2	Paragraph/clause 1·1·2[30] * applies/does not apply
	Maximum liability *(Unless paragraph/clause 1·1·2 is stated to apply and the maximum liability is stated, paragraph/clause 1·1·2 shall not apply.)*	The maximum liability is £ _____
	Type of maximum liability *(If not stated, it shall be an aggregate limit on liability.)*	* Maximum liability is in respect of each breach/ * Maximum liability is an aggregate limit on liability
1·3·1	Net Contribution: Consultants *(If none are specified, these shall be the Architect/Contract Administrator and the Quantity Surveyor (including any replacements), together with any other consultants who agree to give third party rights or collateral warranties (or undertakings in similar terms) to any Purchaser(s) and/or Tenant(s).)*	For the purposes of paragraph/clause 1·3·1[30] 'the Consultants' are: _____ _____ _____ _____
1·3·2	Net Contribution: Sub-Contractors *(If none are specified, these shall be such as agree to give third party rights or collateral warranties (or undertakings in similar terms) to any Purchaser(s) and/or Tenant(s).)*	For the purposes of paragraph/clause 1·3·2[30] 'the Sub-Contractors' are: _____ _____ _____ _____

Funder Rights Particulars

Clauses 7B, 7D and 7E of the Conditions	**(C) Identity of Funder in whom Funder Rights may be vested under this Contract** *(If not identified by name, class or description either here or in a document identified at the commencement of Part 2, Funder Rights shall not be required from the Contractor.)*	_____ _____

---

[30]   The paragraph numbers in Schedule 5 are the same as the clause numbers in the JCT Collateral Warranty.

(D) Funder Rights from the Contractor

Nature of Funder Rights from the Contractor
*(If neither clause reference is deleted, clause 7B
shall apply.)*

*   Clause 7B (Third Party Rights) applies/
*   Clause 7D (Collateral Warranty) applies

1·1   Net Contribution: Consultants and Sub-
Contractors
*(Unless otherwise stated, these shall be those
specified (or deemed to be specified) under (B)
above.)*

    _____

    _____

6·3   Period required, if other than 7 days

    _____

Collateral Warranties from Sub-Contractors

(E) If warranties are required from sub-contractors, set out the necessary particulars in a separate document and state here the reference number, date and/or other identifier of that document.[27][31][32][33]

_____

_____

If or to the extent not included in any such document complete the particulars below:

Clauses 3·7 and 3·9 of the Conditions	Sub-contractors from whom Warranties may be required[31]	Type(s) of warranty required[32] (SCWa/P&T, SCWa/F, SCWa/E) limited to	Level of Professional Indemnity insurance required (if applicable)[33]
	_____	_____	_____

For these purposes and for the purposes of any document identified above, unless otherwise stated:

(i)   all Purchasers and Tenants identified at (A) above, any Funder identified at (C) above and the Employer shall be entitled to Collateral Warranties in accordance with clause 7E from each identified sub-contractor[32];

(ii)   **if applicable, the level of Professional Indemnity insurance must be specified**[33]; the basis of that cover shall be whichever applies under the Contract Particulars for clause 6·11;

(iii)   if a maximum liability is specified under (B) above, that shall also apply in relation to all sub-contractors' Collateral Warranties unless a lower amount is specified;

(iv)   "the Consultants" for sub-contractors' Collateral Warranties shall be those stated in (B) above;

(v)   if a period other that 7 days is specified in (D) above, that other period shall apply in clause 6·3 of each Collateral Warranty SCWa/E and SCWa/F.

---

[31]   Employers should be selective in listing the sub-contractors (or categories of sub-contractor) from whom collateral warranties may be required (see the Guide).

[32]   Where a sub-contractor is required to grant only a particular type or types of the Collateral Warranties referred to in clause 7E (i.e. the Sub-Contractor Collateral Warranty for a Purchaser or Tenant (SCWa/P&T), for a Funder (SCWa/F) and for the Employer (SCWa/E)), state the particular type(s). All three Collateral Warranties are documents prepared by JCT.

[33]   Professional Indemnity insurance applies only where the sub-contractor has design responsibilities. As to cover levels, see the Guide.

Note on Execution

This Agreement should be executed by both the Employer and the Contractor either under hand or as a deed. As to factors relevant to that choice, see the Guide.

Execution under hand

If this Agreement is to be executed under hand, use the form set out on the following page. Each Party or his authorised representative should sign where indicated in the presence of a witness who should then sign and set out his name and address.

Execution as a Deed

If this Agreement is to be executed as a deed, each Party should use the relevant form marked 'Execution as a Deed' in accordance with the notes provided.

Other forms of Attestation

In cases where the forms of attestation set out are not appropriate, e.g. in the case of certain housing associations and partnerships or if a Party wishes an attorney to execute this Agreement on his behalf, the appropriate form(s) may be inserted in the vacant space opposite and/or below.

As witness

the hands of the Parties
or their duly authorised representatives

Signed by or on behalf of the
Employer

_____

in the presence of:

_____

witness' signature

_____

witness' name

_____

witness' address

_____

Signed by or on behalf of the
Contractor

_____

in the presence of:

_____

witness' signature

_____

witness' name

_____

witness' address

_____

Notes on Execution as a Deed

1. For the purposes of execution as a deed, two forms are provided for execution, one for the Employer and the other for the Contractor. Each form provides four methods of execution, **(A)** to **(D)**, for use as appropriate. The full name of the Employer or Contractor (whether an individual, a company or other body) should be inserted where indicated at the commencement of the relevant form. This applies irrespective of the method used.

2. For public and private companies incorporated and registered under the Companies Acts, the three principal methods of execution as a deed are:

   (A) through signature by a Director and the Company Secretary or by two Directors;

   (B) by affixing the company's common seal in the presence of a Director and the *Company* Secretary or of two Directors or other duly authorised officers; or

   (C) signature by a single Director in the presence of a witness who attests the signature.

   Methods **(A)** and **(C)** are available to public and private companies whether or not they have a common seal. (Method **(C)** was introduced by section 44(2)(b) of the Companies Act 2006.) Methods **(A)** and **(C)** are not available under companies legislation to local authorities or to certain other bodies corporate, e.g. bodies incorporated by letters patent or private Act of Parliament that are not registered under companies legislation and such bodies may only use method **(B)**.

3. Where method **(A)** is being used, delete the inappropriate words and insert in the spaces indicated the names of the two Directors, or of the Director and the Company Secretary, who are to sign.

4. If method **(B)** (affixing the common seal) is adopted in cases where either or both the authorised officers attesting its affixation are not themselves a Director or the *Company* Secretary, their respective office(s) should be substituted for the reference(s) to Director and/or to *Company* Secretary/Director. (In the case of execution by bodies that are not companies, the reference to "*Company*" under the second signature should be deleted where appropriate.)

5. Method **(C)** (execution by a single Director) has been introduced primarily, but not exclusively, for 'single officer' companies. The Director should sign where indicated in the presence of a witness who should then sign and set out his name and address.

6. Where the Employer or Contractor is an individual, he should use method **(D)** and sign where indicated in the presence of a witness who should then sign and set out his name and address.

Executed as a Deed by the Employer

namely [1] _____

(A) acting by a Director and the Company Secretary/two Directors **of the company** [2, 3]

_____   and   _____
(Print name of signatory)   *(Print name of signatory)*

_____   _____
Signature   Director   *Signature*   Company Secretary/Director

(B) by affixing hereto the common seal **of the company/other body corporate** [2, 4]

in the presence of

_____
Signature   Director

_____
Signature   Company Secretary/Director

[Common seal of company]

(C) by attested signature of a single Director **of the company** [2, 5]

_____
Signature   Director

in the presence of

Witness' signature _____ *(Print name)* _____

Witness' address _____

(D) by attested signature **of the individual** [6]

_____
Signature

in the presence of

Witness' signature _____ *(Print name)* _____

Witness' address _____

Note: The numbers on this page refer to the numbered paragraphs in the Notes on Execution as a Deed.

Executed as a Deed by the Contractor

namely [1] _____

(A) acting by a Director and the Company Secretary/two Directors **of the company** [2, 3]

_____   and   _____
(Print name of signatory)   *(Print name of signatory)*

_____   _____
Signature   Director   *Signature*   Company Secretary/Director

(B) by affixing hereto the common seal **of the company/other body corporate** [2, 4]

in the presence of

_____
Signature   Director

_____
Signature   Company Secretary/Director

[Common seal of company]

(C) by attested signature of a single Director **of the company** [2, 5]

_____
Signature   Director

in the presence of

Witness' signature _____ *(Print name)* _____

Witness' address _____

(D) by attested signature **of the individual** [6]

_____
Signature

in the presence of

Witness' signature _____ *(Print name)* _____

Witness' address _____

Note: The numbers on this page refer to the numbered paragraphs in the Notes on Execution as a Deed.

Conditions

Section 1   Definitions and Interpretation

Definitions

1·1   Unless the context otherwise requires or the Agreement or these Conditions specifically provide otherwise, the following words and phrases, where they appear in capitalised form in the Agreement or these Conditions, shall have the meanings stated or referred to below:

Word or phrase	Meaning
Activity Schedule:	see the **Second Recital**.
Adjudicator:	an individual appointed under **clause 9·2** as the Adjudicator.
Agreement:	the Articles of Agreement to which these Conditions are annexed, consisting of the Recitals, the Articles and the Contract Particulars.
All Risks Insurance:	see **clause 6·8**.
Approximate Quantity:	a quantity in the Contract Bills there identified as an approximate quantity.
Arbitrator:	an individual appointed under **clause 9·4** as the Arbitrator.
Architect/Contract Administrator:	the person named in **Article 3** or any successor nominated or otherwise agreed under **clause 3·5**.
Article:	an article in the **Agreement**.
Base Date:	the date stated as such date in the **Contract Particulars** (against the reference to **clause 1·1**)[34].
Business Day:	any day which is not a Saturday, a Sunday or a Public Holiday.
CDM Co-ordinator:	the Architect/Contract Administrator or other person named in **Article 5** or any successor appointed by the Employer.
CDM Planning Period:	the minimum amount of time referred to in regulation 10(2)(c) of the CDM Regulations, as specified in the **Contract Particulars** (against the reference to **clause 1·1**).
CDM Regulations:	the Construction (Design and Management) Regulations 2007.
CDP Analysis:	see the **Ninth Recital** and the **Contract Particulars**.
CDP Documents:	the Employer's Requirements, the Contractor's Proposals, the CDP Analysis and the further documents referred to in **clause 2·9·2**.
CDP Works:	that part of the Works comprised in the Contractor's Designed Portion.

---

[34]   The Base Date is relevant (inter alia) to clause 2·17·2·1 (changes in Statutory Requirements) and the Fluctuations Options (Schedule 7) and it helps to determine the edition/issue and/or version of documents relevant to this Contract, e.g. the Standard Method of Measurement and definitions of the prime cost of daywork (clause 5·7).

continued 1·1

Certificate of Making Good:	see **clause 2·39**.
Completion Date:	the Date for Completion of the Works or of a Section as stated in the **Contract Particulars** or such other date as is fixed either under **clause 2·28** or by a Pre-agreed Adjustment.
Conditions:	the clauses set out in sections 1 to 9 of these Conditions, together with and including the Schedules hereto.
Confirmed Acceptance:	the Architect/Contract Administrator's confirmation of acceptance of a Schedule 2 Quotation in accordance with paragraph 3·2 of **Schedule 2**.
Construction Industry Scheme (or 'CIS'):	see the **Fourth Recital**.
Construction Phase Plan:	the plan prepared by the Principal Contractor, where the project is notifiable under the CDM Regulations and in order to comply with regulation 23, including any updates and revisions.
Contract Bills:	the fully priced bills of quantities referred to in the **Second Recital**.
Contract Documents:	the Contract Drawings, the Contract Bills, the Agreement and these Conditions, together with (where applicable) the Employer's Requirements, the Contractor's Proposals and the CDP Analysis.
Contract Drawings:	the drawings referred to in the **Third Recital**.
Contract Particulars:	the particulars in the **Agreement** and there described as such, including the entries made by the Parties.
Contract Sum:	the sum stated in **Article 2**.
Contractor:	the person named as Contractor in the **Agreement**.
Contractor's Design Documents:	the drawings, details and specifications of materials, goods and workmanship and other related documents prepared by or for the Contractor in relation to the Contractor's Designed Portion.
Contractor's Designed Portion:	see the **Seventh Recital**.
Contractor's Persons:	the Contractor's employees and agents, all other persons employed or engaged on or in connection with the Works or any part of them and any other person properly on the site in connection therewith, excluding the Architect/Contract Administrator, the Quantity Surveyor, the Employer, Employer's Persons and any Statutory Undertaker.
Contractor's Proposals:	see the **Ninth Recital** and the **Contract Particulars**.
Date for Completion:	the date stated as such date in the **Contract Particulars** (against the reference to **clause 1·1**) in relation to the Works or a Section.
Date of Possession:	the date stated as such date in the **Contract Particulars** (against the reference to **clause 2·4**) in relation to the Works or a Section.
Employer:	the person named as Employer in the **Agreement**.
Employer's Persons:	all persons employed, engaged or authorised by the Employer, excluding the Contractor, Contractor's Persons, the Architect/Contract Administrator, the Quantity Surveyor and any Statutory Undertaker but including any such third party as is referred to in clause 3·23.
Employer's Requirements:	see the **Eighth Recital** and the **Contract Particulars**.

continued 1·1

Excepted Risks:	see **clause 6·8**.
Final Certificate:	see **clauses 1·10** and **4·15**.
Finance Agreement:	the agreement between the Funder and the Employer for the provision of finance for the Works.
Fluctuations Options A, B and C:	the provisions set out in **Schedule 7** (see **clause 4·21** and the **Contract Particulars**).
Funder:	the person named or otherwise sufficiently identified as such in or by the Funder Rights Particulars and in respect of whom the Employer gives notice under **clause 7B·1**.
Funder Rights:	the rights in favour of the Funder set out in **Part 2 of Schedule 5** or in the appropriate form of collateral warranty.
Funder Rights Particulars:	the entries against **clause 6·11** in **Part 1** and the relevant items and entries in that section of **Part 2 of the Contract Particulars**.
Gross Valuation:	see **clause 4·16**.
Information Release Schedule:	the schedule referred to in the **Fifth Recital**.
Insolvent:	see **clause 8·1**.
Insurance Options A, B and C:	the provisions relating to insurance of the Works and (where applicable) existing structures set out in **Schedule 3**.
Interest Rate:	a rate 5% per annum above the official dealing rate of the Bank of England current at the date that a payment due under this Contract becomes overdue.
Interim Certificate:	any of the certificates to which **clause 4·9** refers.
Joint Fire Code:	the Joint Code of Practice on the Protection from Fire of Construction Sites and Buildings Undergoing Renovation, published by the Construction Confederation and the Fire Protection Association with the support of the Association of British Insurers, the Chief and Assistant Chief Fire Officers Association and the London Fire Brigade, as amended/revised from time to time.
Joint Names Policy:	see **clause 6·8**.
Listed Items:	materials, goods and/or items prefabricated for inclusion in the Works which are listed as such items by the Employer in a list supplied to the Contractor and annexed to the Contract Bills.
Non-Completion Certificate:	see **clause 2·31**.
P&T Rights:	the rights in favour of a Purchaser or Tenant set out in **Part 1 of Schedule 5** or in the appropriate form of collateral warranty.
P&T Rights Particulars:	the entries against **clause 6·11** in **Part 1** and the relevant items and entries in that section of **Part 2 of the Contract Particulars**.
Parties:	the Employer and the Contractor together.
Party:	either the Employer or the Contractor.
Practical Completion Certificate:	see **clause 2·30**.

continued 1·1

Pre-agreed Adjustment:	see **clause 2·26·2**.
Principal Contractor:	the Contractor or other contractor named in **Article 6** or any successor appointed by the Employer.
Provisional Sum:	includes a sum provided for work, whether or not identified as being for defined or undefined work within the meaning of General Rule 10 of the Standard Method of Measurement.
Public Holiday:	Christmas Day, Good Friday or a day which under the Banking and Financial Dealings Act 1971 is a bank holiday.[35]
Purchaser:	any person named or otherwise sufficiently identified as such (whether by class or description) in or by the P&T Rights Particulars to whom the Employer transfers or agrees to transfer his interest in all or part of the Works.
Quantity Surveyor:	the person named in **Article 4** or any successor nominated or otherwise agreed under **clause 3·5**.
Recitals:	the recitals in the **Agreement**.
Rectification Period:	the period stated as such period in the **Contract Particulars** (against the reference to **clause 2·38**) in relation to the Works or (where applicable) a Section.
Relevant Date:	see **clause 2·33**.
Relevant Event:	see **clause 2·29**.
Relevant Matter:	see **clause 4·24**.
Relevant Omission:	see **clause 2·26·3**.
Relevant Part:	see **clause 2·33**.
Retention:	see **clauses 4·10** and **4·18** to **4·20**.
Retention Bond:	see **clause 4·19**.
Retention Percentage:	see **clause 4·20** and the **Contract Particulars**.
Schedule 2 Quotation:	see **clause 5·3** and **Schedule 2**.
Scheme:	Part 1 of the Schedule to The Scheme for Construction Contracts (England and Wales) Regulations 1998.
Sections:	(where applicable) the Sections into which the Works have been divided, as referred to in the **Sixth Recital** and the **Contract Particulars**.
Section Completion Certificate:	see **clause 2·30·2**.
Section Sum:	see **clause 2·37** and the **Contract Particulars**.
Site Materials:	all unfixed materials and goods delivered to and placed on or adjacent to the Works which are intended for incorporation therein.
Specified Perils:	see **clause 6·8**.

---

[35] Amend as necessary if different Public Holidays are applicable.

© **The Joint Contracts Tribunal Limited 2007**

continued 1·1

Standard Method of Measurement:	the Standard Method of Measurement of Building Works, 7th Edition, produced by The Royal Institution of Chartered Surveyors and the Construction Confederation, current, unless otherwise stated in the Contract Bills, at the Base Date (references in that publication to 'the Appendix' being read as references to the Contract Particulars).
Statutory Requirements:	any statute, statutory instrument, regulation, rule or order made under any statute or directive having the force of law which affects the Works or performance of any obligations under this Contract and any regulation or bye-law of any local authority or statutory undertaker which has any jurisdiction with regard to the Works or with whose systems the Works are, or are to be, connected.
Statutory Undertaker:	any local authority or statutory undertaker where executing work solely in pursuance of its statutory obligations, including any persons employed, engaged or authorised by it upon or in connection with that work.
Tenant:	any person named or otherwise sufficiently identified as such (whether by class or description) in or by the P&T Rights Particulars to whom the Employer grants or agrees to grant a leasehold interest in all or part of the Works.
Terrorism Cover:	see **clause 6·8**.
Valuation:	a valuation by the Quantity Surveyor in accordance with the Valuation Rules, pursuant to **clause 5·2**.
Valuation Rules:	see **clauses 5·6** to **5·10**.
Variation:	see **clause 5·1**.
VAT:	Value Added Tax.
Works:	the works briefly described in the **First Recital** (including, where applicable, the CDP Works), as more particularly shown, described or referred to in the Contract Documents, including any changes made to those works in accordance with this Contract.

Interpretation

Reference to clauses etc.

1·2 Unless otherwise stated, a reference in the Agreement or in these Conditions to a clause or Schedule is to that clause in or Schedule to these Conditions and, unless the context otherwise requires, a reference in a Schedule to a paragraph is to that paragraph of that Schedule.

Agreement etc. to be read as a whole

1·3 The Agreement and these Conditions are to be read as a whole but nothing contained in the Contract Bills or the CDP Documents shall override or modify the Agreement or these Conditions.

Headings, references to persons, legislation etc.

1·4 In the Agreement and these Conditions, unless the context otherwise requires:

·1 the headings are included for convenience only and shall not affect the interpretation of this Contract;

·2 the singular includes the plural and vice versa;

·3 a gender includes any other gender;

continued 1·4

·4 a reference to a 'person' includes any individual, firm, partnership, company and any other body corporate; and

·5 a reference to a statute, statutory instrument or other subordinate legislation ('legislation') is to such legislation as amended and in force from time to time, including any legislation which re-enacts or consolidates it, with or without modification.

Reckoning periods of days

1·5 Where under this Contract an act is required to be done within a specified period of days after or from a specified date, the period shall begin immediately after that date. Where the period would include a day which is a Public Holiday that day shall be excluded.

Contracts (Rights of Third Parties) Act 1999

1·6 Other than such rights of any Purchasers, Tenants and/or Funder as take effect pursuant to clauses 7A and/or 7B, nothing in this Contract confers or is intended to confer any right to enforce any of its terms on any person who is not a party to it.

Giving or service of notices and other documents

1·7 Subject to the specific provisions of this Contract as to the manner of giving or service of any notice or other document:

·1 any notice or document may be given or served by any effective means and shall be deemed to be duly given or served if addressed and given by actual delivery or sent by pre-paid post to the Party to be served at the address stated in the Contract Particulars or such other address as may from time to time be agreed;

·2 if there is then no current agreed address for the Party to be served, the notice or other document shall be effectively served if given by actual delivery or sent by pre-paid post to the Party's last known principal business address, or, if a body corporate, its registered or principal office.

Electronic communications

1·8 Subject to the specific provisions referred to in clause 1·7, other communications between the Parties may be made in accordance with any procedures stated or identified in the Contract Particulars or otherwise agreed in writing by the Parties.

Issue of Architect/Contract Administrator's certificates

1·9 Except where otherwise specifically provided, each certificate to be issued by the Architect/Contract Administrator under these Conditions shall be issued to the Employer, and immediately upon issue the Architect/Contract Administrator shall send a copy of it to the Contractor.

Effect of Final Certificate

1·10 ·1 Except as provided in clauses 1·10·2, 1·10·3 and 1·10·4 (and save in respect of fraud), the Final Certificate shall have effect in any proceedings under or arising out of or in connection with this Contract (whether by adjudication, arbitration or legal proceedings) as:

·1 conclusive evidence that where and to the extent that any of the particular qualities of any materials or goods or any particular standard of an item of workmanship was described expressly in the Contract Drawings or the Contract Bills, or in any instruction issued by the Architect/Contract Administrator under these Conditions or in any drawings or documents issued by the Architect/Contract Administrator under any of clauses 2·9 to 2·12, to be for the approval of the Architect/Contract Administrator, the particular quality or standard was to the reasonable satisfaction of the Architect/Contract Administrator, but the Final Certificate shall not be conclusive evidence that they or any other materials or goods or workmanship comply with any other requirement or term of this Contract;

continued 1·10·1

·2 conclusive evidence that any necessary effect has been given to all the terms of this Contract which require that an amount be added to or deducted from the Contract Sum or that an adjustment be made to the Contract Sum save where there has been any accidental inclusion or exclusion of any work, materials, goods or figure in any computation or any arithmetical error in any computation, in which event the Final Certificate shall have effect as conclusive evidence as to all other computations;

·3 conclusive evidence that all and only such extensions of time, if any, as are due under clause 2·28 have been given; and

·4 conclusive evidence that the reimbursement of direct loss and/or expense, if any, to the Contractor pursuant to clause 4·23 is in final settlement of all and any claims which the Contractor has or may have arising out of the occurrence of any of the Relevant Matters, whether such claim be for breach of contract, duty of care, statutory duty or otherwise.

·2 If any adjudication, arbitration or other proceedings have been commenced by either Party before the Final Certificate is issued, the Final Certificate shall have effect as conclusive evidence as provided in clause 1·10·1 upon and from the earlier of either:

·1 the conclusion of such proceedings, in which case the Final Certificate shall be subject to the terms of any decision, award or judgment in or settlement of such proceedings; or

·2 the expiry of any period of 12 months from or after the issue of the Final Certificate during which neither Party takes any further step in such proceedings, in which case the Final Certificate shall be subject to any terms agreed in settlement of any of the matters previously in issue in such proceedings.

·3 If any adjudication, arbitration or other proceedings are commenced by either Party within 28 days after the Final Certificate has been issued, the Final Certificate shall have effect as conclusive evidence as provided in clause 1·10·1 save only in respect of the matters to which those proceedings relate.

·4 In the case of a dispute or difference on which an Adjudicator gives his decision on a date which is after the date of issue of the Final Certificate, if either Party wishes to have that dispute or difference determined by arbitration or legal proceedings, that Party may commence arbitration or legal proceedings within 28 days of the date on which the Adjudicator gives his decision.

Effect of certificates other than Final Certificate

1·11 Save as stated in clause 1·10 no certificate of the Architect/Contract Administrator shall of itself be conclusive evidence that any works, any materials or goods or any design completed by the Contractor for the Contractor's Designed Portion to which the certificate relates are in accordance with this Contract.

Applicable law

1·12 This Contract shall be governed by and construed in accordance with the law of England.[36]

---

[36] Where the Parties do not wish the law applicable to this Contract to be the law of England appropriate amendments should be made.

Contractor's Obligations

General obligations

2·1   The Contractor shall carry out and complete the Works in a proper and workmanlike manner and in compliance with the Contract Documents, the Construction Phase Plan and other Statutory Requirements, and shall give all notices required by the Statutory Requirements.

Contractor's Designed Portion

2·2   Where the Works include a Contractor's Designed Portion, the Contractor shall:

·1   in accordance with the Contract Drawings and the Contract Bills (to the extent they are relevant), complete the design for the Contractor's Designed Portion, including the selection of any specifications for the kinds and standards of the materials, goods and workmanship to be used in the CDP Works, so far as not described or stated in the Employer's Requirements or the Contractor's Proposals;

·2   comply with the directions of the Architect/Contract Administrator for the integration of the design of the Contractor's Designed Portion with the design of the Works as a whole, subject to the provisions of clause 3·10·3; and

·3   in complying with this clause 2·2, comply with regulations 11, 12 and 18 of the CDM Regulations.

Materials, goods and workmanship

2·3   ·1   All materials and goods for the Works, excluding any CDP Works, shall, so far as procurable, be of the kinds and standards described in the Contract Bills. Materials and goods for any CDP Works shall, so far as procurable, be of the kinds and standards described in the Employer's Requirements or, if not there specifically described, as described in the Contractor's Proposals or documents referred to in clause 2·9·2. The Contractor shall not substitute any materials or goods so described without the written consent of the Architect/Contract Administrator, which shall not be unreasonably delayed or withheld but shall not relieve the Contractor of his other obligations.

·2   Workmanship for the Works, excluding any CDP Works, shall be of the standards described in the Contract Bills. Workmanship for any CDP Works shall be of the standards described in the Employer's Requirements or, if not there specifically described, as described in the Contractor's Proposals.

·3   Where and to the extent that approval of the quality of materials or goods or of the standards of workmanship is a matter for the opinion of the Architect/Contract Administrator, such quality and standards shall be to his reasonable satisfaction. To the extent that the quality of materials and goods or standards of workmanship are neither described in the manner referred to in clause 2·3·1 or 2·3·2 nor stated to be a matter for such opinion or satisfaction, they shall in the case of the Contractor's Designed Portion be of a standard appropriate to it and shall in any other case be of a standard appropriate to the Works.

·4   The Contractor shall upon the request of the Architect/Contract Administrator provide him with reasonable proof that the materials and goods used comply with this clause 2·3.

·5   The Contractor shall take all reasonable steps to encourage Contractor's Persons to be registered cardholders under the Construction Skills Certification Scheme (CSCS) or qualified under an equivalent recognised qualification scheme.

Possession

Date of Possession – progress

2·4   On the Date of Possession possession of the site or, in the case of a Section, possession of the relevant part of the site shall be given to the Contractor who shall thereupon begin the construction of the Works or Section and regularly and diligently proceed with and complete the same on or before the relevant Completion Date. For the purposes of the Works insurances the Contractor shall retain possession:

·1   of the site and the Works up to and including the date of issue of the Practical Completion Certificate; or

·2   of each Section and the relevant part of the site up to and including the date of issue of the Section Completion Certificate for that Section and, in respect of any balance of the site, up to and including the date of issue of the Practical Completion Certificate

and, subject to clause 2·33 and section 8, the Employer shall not be entitled to take possession of any part or parts of the Works or Section until such date.

Deferment of possession

2·5   If the Contract Particulars state that clause 2·5 applies in respect of the Works or any Section, the Employer may defer the giving of possession of the site or relevant part of it for a period not exceeding 6 weeks or lesser period stated in the Contract Particulars, calculated from the relevant Date of Possession.

Early use by Employer

2·6   ·1   Notwithstanding clause 2·4, the Employer may, with the consent in writing of the Contractor, use or occupy the site or the Works or part of them, whether for storage or otherwise, before the date of issue of the Practical Completion Certificate or relevant Section Completion Certificate. Before the Contractor gives his consent to such use or occupation, the Contractor or the Employer shall notify the insurers under whichever of Insurance Options A, B or C (paragraph C·2) applies and obtain confirmation that such use or occupation will not prejudice the insurance. Subject to such confirmation, the consent of the Contractor shall not be unreasonably delayed or withheld.

·2   Where Insurance Option A applies and the insurers' confirmation is conditional on an additional premium being paid, the Contractor shall notify the Employer of the amount of it. If the Employer continues to require such use or occupation, the additional premium shall be added to the Contract Sum and the Contractor shall if requested produce the receipt for it to the Employer.

Work not forming part of the Contract

2·7   In regard to any work not forming part of this Contract which the Employer requires to be carried out by the Employer himself or by any Employer's Persons:

·1   where the Contract Bills provide the information necessary to enable the Contractor to carry out and complete the Works or each relevant Section in accordance with this Contract, the Contractor shall permit the execution of such work;

·2   where the Contract Bills do not provide the information referred to in clause 2·7·1, the Employer may with the consent of the Contractor arrange for the execution of such work, such consent not to be unreasonably delayed or withheld.

Supply of Documents, Setting Out etc.

Contract Documents

2·8   ·1   The Contract Documents shall remain in the custody of the Employer and shall be available at all reasonable times for inspection by the Contractor.

·2   Immediately after the execution of this Contract the Architect/Contract Administrator, without charge to the Contractor, shall (unless previously provided) provide him with:

continued 2·8·2

  ·1   one copy, certified on behalf of the Employer, of the Contract Documents;

  ·2   two further copies of the Contract Drawings; and

  ·3   two copies of the unpriced bills of quantities.

·3   The Contractor shall keep upon the site and available to the Architect/Contract Administrator or his representative at all reasonable times a copy of each of the following documents, namely: the Contract Drawings; the unpriced bills of quantities; the CDP Documents (where applicable); the descriptive schedules or similar documents referred to in clause 2·9·1·1; the master programme referred to in clause 2·9·1·2; and the drawings and details referred to in clauses 2·10 and 2·12.

·4   None of the documents referred to in this clause 2·8 or provided or released to the Contractor in accordance with clauses 2·9 to 2·12 shall be used by the Contractor for any purpose other than this Contract, and the Employer, the Architect/Contract Administrator and the Quantity Surveyor shall not divulge or use except for the purposes of this Contract any of the rates or prices in the Contract Bills.

Construction information and Contractor's master programme

2·9 ·1   As soon as possible after the execution of this Contract, if not previously provided:

  ·1   the Architect/Contract Administrator, without charge to the Contractor, shall provide him with 2 copies of any descriptive schedules or similar documents necessary for use in carrying out the Works (excluding any CDP Works), together with any pre-construction information required for the purposes of regulation 10 of the CDM Regulations; and

  ·2   the Contractor shall without charge provide the Architect/Contract Administrator with 2 copies of his master programme for the execution of the Works and, within 14 days of any decision by the Architect/Contract Administrator under clause 2·28·1 or of agreement of any Pre-agreed Adjustment, with 2 copies of an amendment or revision of that programme to take account of that decision or agreement,

but nothing in the descriptive schedules or similar documents (or in that master programme or any amendment or revision of it) shall impose any obligation beyond those imposed by the Contract Documents.

·2   In relation to any CDP Works, the Contractor, in addition to complying with regulations 11, 12 and 18 of the CDM Regulations, shall without charge provide the Architect/Contract Administrator with 2 copies of:

  ·1   such Contractor's Design Documents, and (if requested) related calculations and information, as are reasonably necessary to explain or amplify the Contractor's Proposals; and

  ·2   all levels and setting out dimensions which the Contractor prepares or uses for the purposes of carrying out and completing the Contractor's Designed Portion.

·3   The Contractor's Design Documents and other information referred to in clause 2·9·2·1 shall be provided to the Architect/Contract Administrator as and when necessary from time to time in accordance with the Contractor's Design Submission Procedure set out in Schedule 1 or as otherwise stated in the Contract Documents, and the Contractor shall not commence any work to which such a document relates before that procedure has been complied with.

Levels and setting out of the Works

2·10   The Architect/Contract Administrator shall determine any levels required for the execution of the Works and, subject to clause 2·9·2·2, shall provide the Contractor by way of accurately dimensioned drawings with such information as shall enable the Contractor to set out the Works. The Contractor shall be responsible for, and shall at no cost to the Employer amend, any errors arising from his own inaccurate setting out. With the consent of the Employer the Architect/Contract Administrator may instruct that such errors shall not be amended and an appropriate deduction shall be made from the Contract Sum for those that are not required to be amended.

Information Release Schedule

2·11 Except to the extent that the Architect/Contract Administrator is prevented by an act or default of the Contractor or of any of the Contractor's Persons, the Architect/Contract Administrator shall ensure that 2 copies of the information referred to in the Information Release Schedule are released at the time stated in that Schedule. The Employer and the Contractor may agree to vary any such time, such agreement not to be unreasonably withheld.

Further drawings, details and instructions

2·12 ·1 Where not included in the Information Release Schedule, the Architect/Contract Administrator shall from time to time, without charge to the Contractor, provide him with 2 copies of such further drawings or details as are reasonably necessary to explain and amplify the Contract Drawings and shall issue such instructions (including those for or in regard to the expenditure of Provisional Sums) as are necessary to enable the Contractor to carry out and complete the Works in accordance with this Contract.

·2 The further drawings, details and instructions shall be provided or given at the time it is reasonably necessary for the Contractor to receive them, having regard to the progress of the Works, or, if in the Architect/Contract Administrator's opinion practical completion of the Works or relevant Section is likely to be achieved before the relevant Completion Date, having regard to that Completion Date.

·3 Where the Contractor has reason to believe that the Architect/Contract Administrator is not aware of the time by which the Contractor needs to receive such further drawings, details or instructions, he shall, so far as reasonably practicable, advise the Architect/Contract Administrator sufficiently in advance as to enable the Architect/Contract Administrator to comply with this clause 2·12.

Errors, Discrepancies and Divergences

Preparation of Contract Bills and Employer's Requirements

2·13 ·1 Unless in respect of any specified item or items it is otherwise specifically stated in the Contract Bills, the Contract Bills are to have been prepared in accordance with the Standard Method of Measurement and any addendum bills to be issued for the purposes of obtaining a Schedule 2 Quotation shall be prepared on the same basis.

·2 Subject to clause 2·17, the Contractor shall not be responsible for the contents of the Employer's Requirements or for verifying the adequacy of any design contained within them.

Contract Bills and CDP Documents – errors and inadequacy

2·14 ·1 If in the Contract Bills, or any such addendum bill as is referred to in clause 2·13·1, there is any unstated departure from the method of preparation referred to in that clause or any error in description or in quantity or any omission of items (including any error in or omission of information in any item which is the subject of a Provisional Sum for defined work), the departure, error or omission shall not vitiate this Contract but shall be corrected. Where the description of a Provisional Sum for defined work does not provide the information required by the Standard Method of Measurement, the description shall be corrected so that it does provide that information.

·2 If an inadequacy is found in any design in the Employer's Requirements in relation to which the Contractor under clause 2·13·2 is not responsible for verifying its adequacy, then, if or to the extent that the inadequacy is not dealt with in the Contractor's Proposals, the Employer's Requirements shall be altered or modified accordingly.

·3 Subject to clause 2·17, any correction, alteration or modification under clause 2·14·1 or 2·14·2 shall be treated as a Variation.

·4 Any error in description or in quantity in the Contractor's Proposals or in the CDP Analysis or any error consisting of an omission of items from them shall be corrected, but there shall be no addition to the Contract Sum in respect of that correction or in respect of any instruction requiring a Variation of work not comprised in the Contractor's Designed Portion that is necessitated by any such error or its correction.

Notification of discrepancies etc.

2·15 If the Contractor becomes aware of any such departure, error, omission or inadequacy as is referred to in clause 2·14 or any other discrepancy or divergence in or between any of the following documents, namely:

·1 the Contract Drawings;

·2 the Contract Bills;

·3 any instruction issued by the Architect/Contract Administrator under these Conditions;

·4 any drawings or documents issued by the Architect/Contract Administrator under any of clauses 2·9 to 2·12; and

·5 (where applicable) the CDP Documents,

he shall immediately give written notice with appropriate details to the Architect/Contract Administrator, who shall issue instructions in that regard.

Discrepancies in CDP Documents

2·16 ·1 Where the discrepancy or divergence to be notified under clause 2·15 is within or between the CDP Documents other than the Employer's Requirements, the Contractor shall send with his notice, or as soon thereafter as is reasonably practicable, a statement setting out his proposed amendments to remove it. The Architect/Contract Administrator shall not be obliged to issue instructions until he receives that statement, but, when issued, the Contractor shall comply with those instructions and, to the extent that they relate to the removal of that discrepancy or divergence, there shall be no addition to the Contract Sum.

·2 Where the discrepancy is within the Employer's Requirements (including any Variation of them issued under clause 3·14) the Contractor's Proposals shall prevail (subject to compliance with Statutory Requirements), without any adjustment of the Contract Sum. Where the Contractor's Proposals do not deal with such a discrepancy, the Contractor shall inform the Architect/Contract Administrator in writing of his proposed amendment to deal with it and the Architect/Contract Administrator shall either agree the proposed amendment or decide how the discrepancy shall be dealt with; that agreement or decision shall be notified in writing to the Contractor and treated as a Variation.

Divergences from Statutory Requirements

2·17 ·1 If the Contractor or Architect/Contract Administrator becomes aware of any divergence between the Statutory Requirements and any of the documents referred to in clause 2·15, he shall immediately give the other written notice specifying the divergence and, where it is between the Statutory Requirements and any of the CDP Documents, the Contractor shall inform the Architect/Contract Administrator in writing of his proposed amendment for removing it.

·2 Within 7 days of becoming aware of such divergence (or, where applicable, within 14 days of receipt of the Contractor's proposed amendment), the Architect/Contract Administrator shall issue instructions in that regard, in relation to which:

·1 in the case of a divergence between the Statutory Requirements and any of the CDP Documents, the Contractor shall comply at no cost to the Employer unless after the Base Date there is a change in the Statutory Requirements which necessitates an alteration or modification to the Contractor's Designed Portion, in which event such alteration or modification shall be treated as an instruction requiring a Variation of the Employer's Requirements; and

·2 in any other case, if and insofar as those instructions require the Works to be varied, they shall be treated as instructions requiring a Variation.

·3 Provided the Contractor is not in breach of clause 2·17·1, the Contractor shall not be liable under this Contract if the Works (other than the CDP Works) do not comply with the Statutory Requirements to the extent that the non-compliance results from the Contractor having carried out work in accordance with the documents referred to in clauses 2·15·1 to 2·15·4 (other than an instruction for a Variation in respect of the Contractor's Designed Portion).

Emergency compliance with Statutory Requirements

2·18 ·1 If in any emergency compliance with the Statutory Requirements necessitates the Contractor supplying materials and/or executing work before receiving instructions under clause 2·17·2, the Contractor shall supply such limited materials and execute such limited work as are reasonably necessary to secure immediate compliance.

·2 The Contractor shall forthwith inform the Architect/Contract Administrator of the emergency and of the steps that he is taking under clause 2·18·1.

·3 Where the emergency has arisen because of a divergence between the Statutory Requirements and any of the documents referred to in clauses 2·15·1 to 2·15·4, then, provided that the Contractor has complied with clause 2·18·2, work executed and materials supplied by the Contractor under clause 2·18·1 shall be treated as executed and supplied pursuant to an instruction requiring a Variation.

CDP Design Work

Design liabilities and limitation

2·19 Where there is a Contractor's Designed Portion:

·1 insofar as its design is comprised in the Contractor's Proposals and in what the Contractor is to complete in accordance with the Employer's Requirements and these Conditions (including any further design required to be carried out by the Contractor as a result of a Variation), the Contractor shall in respect of any inadequacy in such design have the like liability to the Employer, whether under statute or otherwise, as would an architect or, as the case may be, other appropriate professional designer holding himself out as competent to take on work for such design who, acting independently under a separate contract with the Employer, has supplied such design for or in connection with works to be carried out and completed by a building contractor who is not the supplier of the design.

·2 where and to the extent that this Contract involves the Contractor in taking on work for or in connection with the provision of a dwelling or dwellings, the reference in clause 2·19·1 to the Contractor's liability includes liability under the Defective Premises Act 1972.

·3 where or to the extent that this Contract does not involve the Contractor in taking on work for or in connection with the provision of a dwelling or dwellings to which the Defective Premises Act 1972 applies, the Contractor's liability for loss of use, loss of profit or other consequential loss arising in respect of the liability of the Contractor referred to in clause 2·19·1 shall be limited to the amount, if any, stated in the Contract Particulars; but such limitation of amount shall not apply to or be affected by any liquidated damages which under clause 2·32 the Contractor could be required to pay or allow in the event of failure to complete the Works or a Section by the relevant Completion Date.

Errors and failures – other consequences

2·20 No extension of time shall be given, and clauses 4·23 and 8·9·2 shall not have effect, where or to the extent that the cause of the progress of the Works having been delayed, affected or suspended is:

·1 an error, divergence, omission or discrepancy in the Contractor's Proposals or in anything provided under clause 2·9·2, or any failure of the Contractor, in completing the Contractor's Design Documents, to comply with regulations 11, 12 and 18 of the CDM Regulations; or

·2 failure by the Contractor to provide in due time any necessary Contractor's Design Documents or related calculations or information either:

·1 as required by clause 2·9·3; or

·2 in response to an application in writing from the Architect/Contract Administrator specifying the relevant documents or information and date by which it is reasonably necessary for them to be received, having regard to the progress of the Works (or, where practical completion of the Works or Section is likely to be achieved before the relevant Completion Date, having regard to that Completion Date).

Fees, Royalties and Patent Rights

Fees or charges legally demandable

2·21   The Contractor shall pay all fees or charges (including any rates or taxes) legally demandable under any of the Statutory Requirements and indemnify the Employer against any liability resulting from any failure to do so. Where such fees or charges are stated by way of a Provisional Sum in the Contract Bills they shall be dealt with in accordance with clauses 4·3·2·1 and 4·3·3·3 and in any other case the amount of any such fees or charges (including any rates or taxes other than VAT) shall be added to the Contract Sum unless they:

·1   are priced in the Contract Bills; or

·2   relate solely to the Contractor's Designed Portion (in which case they shall be deemed to be included in the Contract Sum).

Royalties and patent rights – Contractor's indemnity

2·22   All royalties or other sums payable in respect of the supply and use in carrying out the Works as described by or referred to in the Contract Bills or in the Employer's Requirements of any patented articles, processes or inventions shall be deemed to have been included in the Contract Sum, and the Contractor shall indemnify the Employer from and against all claims and proceedings which may be brought or made against the Employer, and all damages, costs and expense to which he may be put, by reason of the Contractor infringing or being held to have infringed any patent rights in relation to any such articles, processes or inventions.

Patent rights – Instructions

2·23   Where in compliance with the Architect/Contract Administrator's instructions the Contractor shall supply and/or use in carrying out the Works any patented articles, processes or inventions, the Contractor shall not be liable in respect of any infringement or alleged infringement of any patent rights in relation to any such articles, processes or inventions and all royalties, damages or other monies which the Contractor may be liable to pay to the persons entitled to such rights shall be added to the Contract Sum.

Unfixed Materials and Goods – property, risk etc.

Materials and goods – on site

2·24   Unfixed materials and goods delivered to, placed on or adjacent to the Works and intended for them shall not be removed except for use on the Works unless the Architect/Contract Administrator has consented in writing to such removal, such consent not to be unreasonably delayed or withheld. Where the value of any such materials or goods has in accordance with clauses 4·10 and 4·16 been included in any Interim Certificate under which the amount properly due to the Contractor has been paid by the Employer, such materials and goods shall become the property of the Employer, but, subject to Insurance Option B or C (if applicable), the Contractor shall remain responsible for loss or damage to them.

Materials and goods – off site

2·25   Where the value of any Listed Items has in accordance with clause 4·17 been included in any Interim Certificate under which the amount properly due to the Contractor has been paid by the Employer, those items shall become the property of the Employer and thereafter the Contractor shall not, except for use upon the Works, remove or cause or permit them to be moved or removed from the premises where they are. The Contractor shall be responsible for any loss of or damage to them and for the cost of their storage, handling and insurance until they are delivered to and placed on or adjacent to the Works. As from such delivery the provisions of clause 2·24 (except the words "Where the value" to "property of the Employer, but,") shall apply to such items.

Adjustment of Completion Date

Related definitions and interpretation

2·26 In clauses 2·27 to 2·29 and, so far as relevant, in the other clauses of these Conditions:

·1 any reference to delay or extension of time includes any further delay or further extension of time;

·2 'Pre-agreed Adjustment' means the fixing of a revised Completion Date for the Works or a Section in respect of a Variation or other work referred to in clause 5·2·1 by the Confirmed Acceptance of a Schedule 2 Quotation;

·3 'Relevant Omission' means the omission of any work or obligation through an instruction for a Variation under clause 3·14 or through an instruction under clause 3·16 in regard to a Provisional Sum for defined work.

Notice by Contractor of delay to progress

2·27 ·1 If and whenever it becomes reasonably apparent that the progress of the Works or any Section is being or is likely to be delayed the Contractor shall forthwith give written notice to the Architect/Contract Administrator of the material circumstances, including the cause or causes of the delay, and shall identify in the notice any event which in his opinion is a Relevant Event.

·2 In respect of each event identified in the notice the Contractor shall, if practicable in such notice or otherwise in writing as soon as possible thereafter, give particulars of its expected effects, including an estimate of any expected delay in the completion of the Works or any Section beyond the relevant Completion Date.

·3 The Contractor shall forthwith notify the Architect/Contract Administrator in writing of any material change in the estimated delay or in any other particulars and supply such further information as the Architect/Contract Administrator may at any time reasonably require.

Fixing Completion Date

2·28 ·1 If, in the opinion of the Architect/Contract Administrator, on receiving a notice and particulars under clause 2·27:

·1 any of the events which are stated to be a cause of delay is a Relevant Event; and

·2 completion of the Works or of any Section is likely to be delayed thereby beyond the relevant Completion Date,

then, save where these Conditions expressly provide otherwise, the Architect/Contract Administrator shall give an extension of time by fixing such later date as the Completion Date for the Works or Section as he then estimates to be fair and reasonable.

·2 Whether or not an extension is given, the Architect/Contract Administrator shall notify the Contractor in writing of his decision in respect of any notice under clause 2·27 as soon as is reasonably practicable and in any event within 12 weeks of receipt of the required particulars. Where the period from receipt to the Completion Date is less than 12 weeks, he shall endeavour to do so prior to the Completion Date.

·3 The Architect/Contract Administrator shall in his decision state:

·1 the extension of time that he has attributed to each Relevant Event; and

·2 (in the case of a decision under clause 2·28·4 or 2·28·5) the reduction in time that he has attributed to each Relevant Omission.

·4 After the first fixing of a later Completion Date in respect of the Works or a Section, either under clause 2·28·1 or by a Pre-agreed Adjustment, but subject to clauses 2·28·6·3 and 2·28·6·4, the Architect/Contract Administrator may by notice in writing to the Contractor, giving the details referred to in clause 2·28·3, fix a Completion Date for the Works or that Section earlier than that previously so *continued 2·28·4* fixed if in his opinion the fixing of such earlier Completion Date is fair and reasonable, having regard to any Relevant Omissions for which instructions have been issued after the last occasion on which a new Completion Date was fixed for the Works or for that Section.

·5 After the Completion Date for the Works or for a Section, if this occurs before the date of practical completion, the Architect/Contract Administrator may, and not later than the expiry of 12 weeks after the date of practical completion shall, by notice in writing to the Contractor, giving the details referred to in clause 2·28·3:

·1 fix a Completion Date for the Works or for the Section later than that previously fixed if in his opinion that is fair and reasonable having regard to any Relevant Events, whether on reviewing a previous decision or otherwise and whether or not the Relevant Event has been specifically notified by the Contractor under clause 2·27·1; or

·2 subject to clauses 2·28·6·3 and 2·28·6·4, fix a Completion Date earlier than that previously fixed if in his opinion that is fair and reasonable having regard to any instructions for Relevant Omissions issued after the last occasion on which a new Completion Date was fixed for the Works or Section; or

·3 confirm the Completion Date previously fixed.

·6 Provided always that:

·1 the Contractor shall constantly use his best endeavours to prevent delay in the progress of the Works or any Section, however caused, and to prevent the completion of the Works or Section being delayed or further delayed beyond the relevant Completion Date;

·2 in the event of any delay the Contractor shall do all that may reasonably be required to the satisfaction of the Architect/Contract Administrator to proceed with the Works or Section;

·3 no decision of the Architect/Contract Administrator under clause 2·28·4 or 2·28·5·2 shall fix a Completion Date for the Works or any Section earlier than the relevant Date for Completion; and

·4 no decision under clause 2·28·4 or 2·28·5·2 shall alter the length of any Pre-agreed Adjustment unless the relevant Variation or other work referred to in clause 5·2·1 is itself the subject of a Relevant Omission.

Relevant Events

2·29 The following are the Relevant Events referred to in clauses 2·27 and 2·28:

·1 Variations and any other matters or instructions which under these Conditions are to be treated as, or as requiring, a Variation;

·2 instructions of the Architect/Contract Administrator:

·1 under any of clauses 2·15, 3·15, 3·16 (excluding an instruction for expenditure of a Provisional Sum for defined work), 3·23 or 5·3·2; or

·2 for the opening up for inspection or testing of any work, materials or goods under clause 3·17 or 3·18·4 (including making good), unless the inspection or test shows that the work, materials or goods are not in accordance with this Contract;

·3 deferment of the giving of possession of the site or any Section under clause 2·5;

·4 the execution of work for which an Approximate Quantity is not a reasonably accurate forecast of the quantity of work required;

·5 suspension by the Contractor under clause 4·14 of the performance of his obligations under this Contract;

·6 any impediment, prevention or default, whether by act or omission, by the Employer, the Architect/ Contract Administrator, the Quantity Surveyor or any of the Employer's Persons, except to the extent caused or contributed to by any default, whether by act or omission, of the Contractor or of any of the Contractor's Persons;

·7 the carrying out by a Statutory Undertaker of work in pursuance of its statutory obligations in relation to the Works, or the failure to carry out such work;

·8 exceptionally adverse weather conditions;

·9 loss or damage occasioned by any of the Specified Perils;

·10 civil commotion or the use or threat of terrorism and/or the activities of the relevant authorities in dealing with such event or threat;

·11 strike, lock-out or local combination of workmen affecting any of the trades employed upon the Works or any of the trades engaged in the preparation, manufacture or transportation of any of the goods or materials required for the Works or any persons engaged in the preparation of the design for the Contractor's Designed Portion;

·12 the exercise after the Base Date by the United Kingdom Government of any statutory power which directly affects the execution of the Works;

·13 force majeure.

Practical Completion, Lateness and Liquidated Damages

Practical completion and certificates

2·30 When in the opinion of the Architect/Contract Administrator practical completion of the Works or a Section is achieved and the Contractor has complied sufficiently with clauses 2·40 and 3·25·4, then:

·1 in the case of the Works, the Architect/Contract Administrator shall forthwith issue a certificate to that effect ('the Practical Completion Certificate');

·2 in the case of a Section, he shall forthwith issue a certificate of practical completion of that Section (a 'Section Completion Certificate');

and practical completion of the Works or the Section shall be deemed for all the purposes of this Contract to have taken place on the date stated in that certificate.

Non-Completion Certificates

2·31 If the Contractor fails to complete the Works or a Section by the relevant Completion Date, the Architect/ Contract Administrator shall issue a certificate to that effect (a 'Non-Completion Certificate'). If a new Completion Date is fixed after the issue of such a certificate, such fixing shall cancel that certificate and the Architect/Contract Administrator shall where necessary issue a further certificate.

Payment or allowance of liquidated damages

2·32 ·1 Provided:

·1 the Architect/Contract Administrator has issued a Non-Completion Certificate for the Works or a Section; and

·2 the Employer has informed the Contractor in writing before the date of the Final Certificate that he may require payment of, or may withhold or deduct, liquidated damages,

the Employer may, not later than 5 days before the final date for payment of the debt due under the Final Certificate, give notice in writing to the Contractor in the terms set out in clause 2·32·2.

continued 2·32

·2 A notice from the Employer under clause 2·32·1 shall state that for the period between the Completion Date and the date of practical completion of the Works or that Section:

·1 he requires the Contractor to pay liquidated damages at the rate stated in the Contract Particulars, or lesser rate stated in the notice, in which event the Employer may recover the same as a debt; and/or

·2 that he will withhold or deduct liquidated damages at the rate stated in the Contract Particulars, or at such lesser stated rate, from monies due to the Contractor.[37]

·3 If the Architect/Contract Administrator fixes a later Completion Date for the Works or a Section or such later Completion Date is stated in the Confirmed Acceptance of a Schedule 2 Quotation, the Employer shall pay or repay to the Contractor any amounts recovered, allowed or paid under clause 2·32 for the period up to that later Completion Date.

·4 If the Employer in relation to the Works or a Section has informed the Contractor in writing in accordance with clause 2·32·1·2 that he may require payment of, or may withhold or deduct, liquidated damages, then, unless the Employer states otherwise in writing, clause 2·32·1·2 shall remain satisfied in relation to the Works or Section, notwithstanding the cancellation of the relevant Non-Completion Certificate and issue of any further Non-Completion Certificate.

Partial Possession by Employer

Contractor's consent

2·33 If at any time or times before the date of issue by the Architect/Contract Administrator of the Practical Completion Certificate or relevant Section Completion Certificate the Employer wishes to take possession of any part or parts of the Works or a Section and the consent of the Contractor has been obtained (which consent shall not be unreasonably delayed or withheld), then, notwithstanding anything expressed or implied elsewhere in this Contract, the Employer may take possession of such part or parts. The Architect/Contract Administrator shall thereupon issue to the Contractor on behalf of the Employer a written statement identifying the part or parts taken into possession and giving the date when the Employer took possession ('the Relevant Part' and 'the Relevant Date' respectively).

Practical completion date

2·34 For the purposes of clauses 2·38 and 4·20·2, practical completion of the Relevant Part shall be deemed to have occurred, and the Rectification Period in respect of the Relevant Part shall be deemed to have commenced, on the Relevant Date.

Defects etc. – Relevant Part

2·35 When in the opinion of the Architect/Contract Administrator any defects, shrinkages or other faults in the Relevant Part which he has required to be made good under clause 2·38 have been made good, he shall issue a certificate to that effect.

Insurance – Relevant Part

2·36 As from the Relevant Date the insurance obligation of the Contractor under Insurance Option A or of the Employer under Insurance Option B or paragraph C·2 of Insurance Option C (whichever applies) shall terminate in respect of the Relevant Part (but not otherwise); and, where Insurance Option C applies, the obligation of the Employer to insure under paragraph C·1 shall from the Relevant Date include the Relevant Part.

Liquidated damages – Relevant Part

2·37 As from the Relevant Date, the rate of liquidated damages stated in the Contract Particulars in respect of the Works or Section containing the Relevant Part shall reduce by the same proportion as the value of the

---

[37] Where the Employer intends to withhold all or any of the liquidated damages payable, either the notice under clause 2·32·2 must comply with the requirements of clause 4·13·4 or 4·15·4 or a separate notice that complies with those requirements must be given.

Relevant Part bears to the Contract Sum or to the relevant Section Sum, as shown in the Contract Particulars.

Defects

Schedules of defects and instructions

2·38   If any defects, shrinkages or other faults in the Works or a Section appear within the relevant Rectification Period due to materials, goods or workmanship not in accordance with this Contract or any failure of the Contractor to comply with his obligations in respect of the Contractor's Designed Portion:

·1   such defects, shrinkages and other faults shall be specified by the Architect/Contract Administrator in a schedule of defects which he shall deliver to the Contractor as an instruction not later than 14 days after the expiry of that Rectification Period; and

·2   notwithstanding clause 2·38·1, the Architect/Contract Administrator may whenever he considers it necessary issue instructions requiring any such defect, shrinkage or other fault to be made good, provided no instructions under this clause 2·38·2 shall be issued after delivery of a schedule of defects or more than 14 days after the expiry of the relevant Rectification Period.

Within a reasonable time after receipt of such schedule or instructions, the defects, shrinkages and other faults shall at no cost to the Employer be made good by the Contractor unless the Architect/Contract Administrator with the consent of the Employer shall otherwise instruct. If he does so otherwise instruct, an appropriate deduction shall be made from the Contract Sum in respect of the defects, shrinkages or other faults not made good.

Certificate of Making Good

2·39   When in the Architect/Contract Administrator's opinion the defects, shrinkages or other faults in the Works or a Section which he has required to be made good under clause 2·38 have been made good, he shall issue a certificate to that effect (a 'Certificate of Making Good'), and completion of that making good shall for the purposes of this Contract be deemed to have taken place on the date stated in that certificate.

Contractor's Design Documents

As-built Drawings

2·40   Where there is a Contractor's Designed Portion, the Contractor, before practical completion of the Works or relevant Section, shall without further charge to the Employer supply for the retention and use of the Employer such Contractor's Design Documents and related information as may be specified in the Contract Documents or as the Employer may reasonably require, showing or describing the Contractor's Designed Portion as built and, without affecting the Contractor's obligations under clause 3·25 (*the health and safety file*), concerning the maintenance and operation of that portion, including any installations forming part of it.

Copyright and use

2·41   ·1   Subject to any rights in any designs, drawings and other documents supplied to the Contractor for the purposes of this Contract by or on behalf of the Employer or the Architect/Contract Administrator, the copyright in all Contractor's Design Documents shall remain vested in the Contractor.

·2   Subject to all monies due and payable under this Contract to the Contractor having been paid, the Employer shall have an irrevocable, royalty-free, non-exclusive licence to copy and use the Contractor's Design Documents and to reproduce the designs and content of them for any purpose relating to the Works including, without limitation, the construction, completion, maintenance, letting, sale, promotion, advertisement, reinstatement, refurbishment and repair of the Works. Such licence shall enable the Employer to copy and use the Contractor's Design Documents for the extension of the Works but shall not include a licence to reproduce the designs contained in them for any extension of the Works.

continued 2·41

·3 The Contractor shall not be liable for any use by the Employer of any of the Contractor's Design Documents for any purpose other than that for which they were prepared.

Access and Representatives

Access for Architect/Contract Administrator

3·1   The Architect/Contract Administrator and any person authorised by him shall at all reasonable times have access to the Works and to the workshops or other premises of the Contractor where work is being prepared for this Contract. When work is to be prepared in workshops or other premises of a sub-contractor the Contractor shall by a term in the sub-contract secure so far as possible a similar right of access to those workshops or premises for the Architect/Contract Administrator and any person authorised by him and shall do all things reasonably necessary to make that right effective. Access under this clause 3·1 may be subject to such reasonable restrictions as are necessary to protect proprietary rights.

Person-in-charge

3·2   The Contractor shall ensure that at all times he has on the site a competent person-in-charge and any instructions given to that person by the Architect/Contract Administrator or directions given to him by the clerk of works in accordance with clause 3·4 shall be deemed to have been issued to the Contractor.

Employer's representative

3·3   The Employer may appoint an individual to act as his representative by giving written notice to the Contractor that from the date stated the individual identified in the notice will exercise all the functions ascribed to the Employer in these Conditions, subject to any exceptions stated in the notice. The Employer may by written notice to the Contractor terminate any such appointment and/or appoint a replacement.[38]

Clerk of works

3·4   The Employer shall be entitled to appoint a clerk of works whose duty shall be to act solely as inspector on behalf of the Employer under the directions of the Architect/Contract Administrator and the Contractor shall afford every reasonable facility for the performance of that duty. If any direction is given to the Contractor by the clerk of works, it shall be of no effect unless given in regard to a matter in respect of which the Architect/Contract Administrator is expressly empowered by these Conditions to issue instructions and unless confirmed in writing by the Architect/Contract Administrator within 2 working days of the direction being given. Any direction so given and confirmed shall, as from the date of issue of that confirmation, be deemed an instruction of the Architect/Contract Administrator.

Replacement of Architect/Contract Administrator or Quantity Surveyor

3·5   ·1   If the Architect/Contract Administrator or Quantity Surveyor at any time ceases to hold that post for the purposes of this Contract, the Employer shall as soon as reasonably practicable, and in any event within 21 days of the cessation, nominate and give the Contractor notice of the identity of a replacement. Except where the Employer is a Local Authority and the nominated replacement is an official of it, the Contractor may within 7 days of the notice give a counter-notice that he objects to the nominated person acting as replacement and, if the Contractor's reasons for doing so are accepted by the Employer or considered sufficient by a person appointed under the dispute resolution procedures of this Contract, the Employer shall withdraw the nomination and nominate an acceptable replacement.

·2   No replacement Architect/Contract Administrator appointed for this Contract shall be entitled to disregard or overrule any certificate, opinion, decision, approval or instruction given or expressed by any predecessor in that post, save to the extent that that predecessor if still in the post would then have had power under this Contract to do so.

---

[38]   To avoid any confusion between the quite distinct roles of the Architect/Contract Administrator and the Quantity Surveyor on the one hand and that of the Employer's representative on the other, neither the Architect/Contract Administrator nor the Quantity Surveyor should be appointed as the Employer's representative.

Contractor's responsibility

3·6 Notwithstanding any obligation of the Architect/Contract Administrator to the Employer and whether or not the Employer appoints a clerk of works, the Contractor shall remain wholly responsible for carrying out and completing the Works in all respects in accordance with these Conditions. That responsibility shall not be affected by the Architect/Contract Administrator or the clerk of works at any time going onto or carrying out any inspection of the Works or visiting any workshop or other premises to inspect them or any work in preparation there, or by the Architect/Contract Administrator including the value of any work, materials or goods in a certificate for payment or by his issuing the Practical Completion Certificate, any Section Completion Certificate or any Certificate of Making Good.

Sub-Letting

Consent to sub-letting

3·7 ·1 The Contractor shall not without the written consent of the Architect/Contract Administrator sub-let the whole or any part of the Works. Such consent shall not be unreasonably delayed or withheld but the Contractor shall remain wholly responsible for carrying out and completing the Works in all respects in accordance with clause 2·1 notwithstanding any such sub-letting.

·2 Where there is a Contractor's Designed Portion, the Contractor shall not without the written consent of the Employer sub-let the design for it. Such consent shall not be unreasonably delayed or withheld but shall not in any way affect the obligations of the Contractor under clauses 2·2 and 2·19 or any other provision of this Contract.

·3 The provisions of this clause 3·7 and of clauses 3·8 and 3·9 shall not apply to the execution of part of the Works by a Statutory Undertaker, who shall not in that capacity be a sub-contractor within the terms of this Contract.

List in Contract Bills

3·8 ·1 Where the Contract Bills provide that certain work measured or otherwise described in those Bills and priced by the Contractor is to be carried out by persons named in a list in or annexed to the Contract Bills and selected from that list by and at the sole discretion of the Contractor, the provisions of this clause 3·8 shall apply.

·2 The list shall comprise not less than three persons. The Employer (or the Architect/Contract Administrator on his behalf) and the Contractor shall each be entitled with the consent of the other, which shall not be unreasonably delayed or withheld, to add additional persons to the list at any time prior to the execution of a binding sub-contract.[39]

·3 If at any time prior to the execution of a binding sub-contract there are less than three persons named in the list that are able and willing to carry out the relevant work, then, either:

·1 the Employer and the Contractor shall by agreement (which shall not be unreasonably delayed or withheld) add the names of other persons so that the list comprises not less than three such persons[39]; or

·2 the work shall be carried out by the Contractor who may sub-let to any sub-contractor in accordance with clause 3·7.

·4 A person selected from the list by the Contractor under this clause 3·8 shall be a sub-contractor.

Conditions of sub-letting

3·9 It shall be a condition of any sub-letting to which clause 3·7 or 3·8 applies that[40]:

---

[39] Any such addition should be confirmed in writing.
[40] The requirements of clauses 3·9·1 and 3·9·2 together with those in paragraphs A·3 and B·4 of the Fluctuations Options (Schedule 7) are met by the JCT Standard Building Sub-Contracts (SBCSub and SBCSub/D).

continued 3·9

·1   the employment of the sub-contractor under the sub-contract shall terminate immediately upon the termination (for any reason) of the Contractor's employment under this Contract;

·2   the sub-contract shall provide:

    ·1   that no unfixed materials and goods delivered to, placed on or adjacent to the Works by the sub-contractor and intended for them shall be removed, except for use on the Works, unless the Contractor has consented in writing to such removal (such consent not to be unreasonably delayed or withheld) and that:

        ·1   where, in accordance with clauses 4·10 and 4·16 of these Conditions, the value of any such materials or goods has been included in any Interim Certificate under which the amount properly due to the Contractor has been paid to him, those materials or goods shall be and become the property of the Employer and the sub-contractor shall not deny that they are and have become the property of the Employer;

        ·2   if the Contractor pays the sub-contractor for any such materials or goods before their value is included in any Interim Certificate, such materials or goods shall upon such payment by the Contractor be and become the property of the Contractor;

    ·2   for the grant by the sub-contractor of the rights of access to workshops or other premises referred to in clause 3·1 of these Conditions;

    ·3   that each party undertakes to the other in relation to the Works and the site duly to comply with the CDM Regulations;

    ·4   that if by the final date for payment stated in the sub-contract the Contractor fails properly to pay any amount, or any part of it, due to the sub-contractor, the Contractor shall in addition to the amount not properly paid pay simple interest thereon at the Interest Rate for the period until such payment is made; such payment of interest to be on and subject to terms equivalent to those of clauses 4·13·6 and 4·15·6 of these Conditions;

    ·5   where applicable, for the execution and delivery by the sub-contractor, in each case within 14 days of receipt of a written request by the Contractor, of such collateral warranties as comply with the Contract Documents;

    ·6   that neither of the provisions referred to in clauses 3·9·2·1·1 and 3·9·2·1·2 shall operate so as to affect any vesting in the Contractor of property in any Listed Item required for the purposes of clause 4·17·2·1 of these Conditions;

·3   the Contractor shall not give such consent as is referred to in clause 3·9·2·1 without the prior consent of the Architect/Contract Administrator under clause 2·24 of these Conditions.

Architect/Contract Administrator's Instructions

Compliance with instructions

3·10   The Contractor shall forthwith comply with all instructions issued to him by the Architect/Contract Administrator in regard to any matter in respect of which the Architect/Contract Administrator is expressly empowered by these Conditions to issue instructions, save that:

·1   where an instruction requires a Variation of the type referred to in clause 5·1·2 the Contractor need not comply to the extent that he makes reasonable objection to it in writing to the Architect/Contract Administrator;

·2   where an instruction for a Variation is given which pursuant to clause 5·3·1 requires the Contractor to provide a Schedule 2 Quotation, the Variation shall not be carried out until the Architect/Contract Administrator has in relation to it issued either a Confirmed Acceptance or a further instruction under clause 5·3·2;

continued 3·10

·3 if in the Contractor's opinion compliance with any direction under clause 2·2·2 or any instruction issued by the Architect/Contract Administrator injuriously affects the efficacy of the design of the Contractor's Designed Portion (including the obligations of the Contractor to comply with regulations 11, 12 and 18 of the CDM Regulations), he shall within 7 days of receipt of the direction or instruction by notice in writing to the Architect/Contract Administrator specify the injurious effect, and the direction or instruction shall not take effect unless confirmed by the Architect/Contract Administrator.

Non-compliance with instructions

3·11   Subject to clause 3·10, if within 7 days after receipt of a written notice from the Architect/Contract Administrator requiring compliance with an instruction the Contractor does not comply, the Employer may employ and pay other persons to execute any work whatsoever which may be necessary to give effect to that instruction. The Contractor shall be liable for all additional costs incurred by the Employer in connection with such employment and an appropriate deduction shall be made from the Contract Sum.

Instructions to be in writing

3·12   ·1   All instructions issued by the Architect/Contract Administrator shall be in writing.

·2   If the Architect/Contract Administrator purports to issue an instruction otherwise than in writing it shall be of no immediate effect but the Contractor shall confirm it in writing to the Architect/Contract Administrator within 7 days, and, if not dissented from in writing to the Contractor within 7 days from receipt of the Contractor's confirmation, it shall take effect as from the expiry of the latter 7 day period.

·3   If within 7 days of giving an instruction otherwise than in writing the Architect/Contract Administrator confirms it in writing, the Contractor shall not be obliged to confirm it and it shall take effect as from the date of the Architect/Contract Administrator's confirmation.

·4   If neither the Contractor nor the Architect/Contract Administrator confirms such an instruction in the manner and time stated but the Contractor nevertheless complies with it, the Architect/Contract Administrator may at any time prior to the issue of the Final Certificate confirm it in writing with retrospective effect.

Provisions empowering instructions

3·13   On receipt of an instruction or purported instruction the Contractor may request the Architect/Contract Administrator to specify in writing which provision of these Conditions empowers its issue and he shall forthwith comply with the request. If the Contractor thereafter complies with that instruction with neither Party then having invoked any dispute resolution procedure under this Contract to establish the Architect/Contract Administrator's powers in that regard, the instruction shall be deemed to have been duly given under the specified provision.

Instructions requiring Variations

3·14   ·1   The Architect/Contract Administrator may issue instructions requiring a Variation.

·2   Any instruction of the type referred to in clause 5·1·2 shall be subject to the Contractor's right of reasonable objection set out in clause 3·10·1.

·3   In respect of the Contractor's Designed Portion, any instruction requiring a Variation shall be an alteration to or modification of the Employer's Requirements.

·4   The Architect/Contract Administrator may sanction in writing any Variation made by the Contractor otherwise than pursuant to an instruction.

·5   No Variation required by the Architect/Contract Administrator or subsequently sanctioned by him shall vitiate this Contract.

Postponement of work

3·15 The Architect/Contract Administrator may issue instructions in regard to the postponement of any work to be executed under this Contract.

Instructions on Provisional Sums

3·16 The Architect/Contract Administrator shall issue instructions in regard to the expenditure of Provisional Sums included in the Contract Bills or in the Employer's Requirements.

Inspection – tests

3·17 The Architect/Contract Administrator may issue instructions requiring the Contractor to open up for inspection any work covered up or to arrange for or carry out any test of any materials or goods (whether or not already incorporated in the Works) or of any executed work. The cost of such opening up or testing (including the cost of making good) shall be added to the Contract Sum unless provided for in the Contract Bills or unless the inspection or test shows that the materials, goods or work are not in accordance with this Contract.

Work not in accordance with the Contract

3·18 If any work, materials or goods are not in accordance with this Contract the Architect/Contract Administrator, in addition to his other powers, may:

·1 issue instructions in regard to the removal from the site of all or any of such work, materials or goods;

·2 after consultation with the Contractor and with the agreement of the Employer, allow all or any of such work, materials or goods to remain (except those which are part of the Contractor's Designed Portion), in which event he shall confirm this in writing to the Contractor but that shall not be construed as a Variation and an appropriate deduction shall be made from the Contract Sum;

·3 after consultation with the Contractor, issue such instructions requiring a Variation as are reasonably necessary as a consequence of any instruction under clause 3·18·1 and/or of any confirmation under clause 3·18·2 (but to the extent that such instructions are reasonably necessary, no addition shall be made to the Contract Sum and no extension of time shall be given); and/or

·4 having due regard to the Code of Practice set out in Schedule 4, issue such instructions under clause 3·17 to open up for inspection or to test as are reasonable in all the circumstances to establish to the reasonable satisfaction of the Architect/Contract Administrator the likelihood or extent, as appropriate to the circumstances, of any further similar non-compliance. To the extent that such instructions are reasonable, whatever the results of the opening up, no addition shall be made to the Contract Sum but clauses 2·28 and 2·29·2·2 shall apply unless the inspection or test shows that the work, materials or goods are not in accordance with this Contract.

Workmanship not in accordance with the Contract

3·19 Where there is any failure to comply with clause 2·1 in regard to the carrying out of work in a proper and workmanlike manner and/or in accordance with the Construction Phase Plan, the Architect/Contract Administrator, in addition to his other powers, may, after consultation with the Contractor, issue such instructions (whether requiring a Variation or otherwise) as are in consequence reasonably necessary. To the extent that such instructions are reasonably necessary, no addition shall be made to the Contract Sum and no extension of time shall be given.

Executed work

3·20 In respect of any materials, goods or workmanship, as comprised in executed work, which under clause 2·3 are to be to the reasonable satisfaction of the Architect/Contract Administrator, the Architect/Contract Administrator, if he is dissatisfied, shall give the reasons for such dissatisfaction to the Contractor within a reasonable time from the execution of the unsatisfactory work.

Exclusion of persons from the Works

3·21   The Architect/Contract Administrator may (but shall not unreasonably or vexatiously) issue instructions requiring the exclusion from the site of any person employed thereon.

Antiquities

Effect of find of antiquities

3·22   All fossils, antiquities and other objects of interest or value which may be found on the site or in excavating it during the progress of the Works shall become the property of the Employer. Upon discovery of any such object the Contractor shall forthwith:

·1   use his best endeavours not to disturb the object and cease work if and insofar as its continuance would endanger the object or prevent or impede its excavation or removal;

·2   take all steps necessary to preserve the object in the exact position and condition in which it was found; and

·3   inform the Architect/Contract Administrator or the clerk of works of its discovery and precise location.

Instructions on antiquities

3·23   The Architect/Contract Administrator shall issue instructions as to action to be taken concerning any object reported under clause 3·22, which (without limiting his powers) may require the Contractor to permit the examination, excavation or removal of the object by a third party.

Loss and expense arising

3·24   If in the Architect/Contract Administrator's opinion compliance with clause 3·22 or with an instruction under clause 3·23 has involved the Contractor in direct loss and/or expense for which he would not be reimbursed under any other provision of this Contract, the Architect/Contract Administrator shall ascertain, or instruct the Quantity Surveyor to ascertain, the amount. Any amount from time to time so ascertained shall be added to the Contract Sum.

CDM Regulations

Undertakings to comply

3·25   Each Party acknowledges that he is aware of and undertakes to the other that in relation to the Works and site he will duly comply with the CDM Regulations. Without limitation, where the project that comprises or includes the Works is notifiable:

·1   the Employer shall ensure both that the CDM Co-ordinator carries out all his duties and, where the Contractor is not the Principal Contractor, that the Principal Contractor carries out all his duties under those regulations;

·2   where the Contractor is and while he remains the Principal Contractor, he shall ensure that:

·1   the Construction Phase Plan is prepared and received by the Employer before construction work under this Contract is commenced, and that any subsequent amendment to it by the Contractor is notified to the Employer, the CDM Co-ordinator and (where not the CDM Co-ordinator) the Architect/Contract Administrator; and

·2   welfare facilities complying with Schedule 2 of the CDM Regulations are provided from the commencement of construction work until the end of the construction phase[41];

---

[41]   There is a duty on contractors to ensure compliance with Schedule 2 of the CDM Regulations so far as is reasonably practicable, whether or not the project is notifiable and whether or not the contractor is the Principal Contractor.

continued 3·25

·3   where the Contractor is not the Principal Contractor, he shall promptly inform the Principal Contractor of the identity of each sub-contractor that he appoints and each sub-subcontractor appointment notified to him;

·4   promptly upon the written request of the CDM Co-ordinator, the Contractor shall provide, and shall ensure that any sub-contractor, through the Contractor, provides, to the CDM Co-ordinator (or, if the Contractor is not the Principal Contractor, to the Principal Contractor) such information as the CDM Co-ordinator reasonably requires for the preparation of the health and safety file.

Appointment of successors

3·26   If the Employer by a further appointment replaces the CDM Co-ordinator or the Principal Contractor, the Employer shall immediately upon such further appointment notify the Contractor in writing of the name and address of the new appointee. If the Employer appoints a successor to the Contractor as the Principal Contractor, the Contractor shall at no cost to the Employer comply with all reasonable requirements of the new Principal Contractor to the extent necessary for compliance with the CDM Regulations; no extension of time shall be given in respect of such compliance.

Contract Sum and Adjustments

Work included in Contract Sum

4·1   The quality and quantity of the work included in the Contract Sum shall be deemed to be that set out in the Contract Bills and, where there is a Contractor's Designed Portion, in the CDP Documents.

Adjustment only under the Conditions

4·2   The Contract Sum shall not be adjusted or altered in any way other than in accordance with the express provisions of these Conditions and, subject to clause 2·14, any error, whether of arithmetic or otherwise, in the computation of the Contract Sum shall be deemed to have been accepted by the Parties.

Items included in adjustments

4·3   ·1   The Contract Sum shall be adjusted by:

  ·1   any amounts agreed by the Employer and the Contractor in respect of Variations, as referred to in clause 5·2·1;

  ·2   the amounts stated in any Schedule 2 Quotations for which the Architect/Contract Administrator has issued to the Contractor a Confirmed Acceptance and by the amount of any Variations thereto as valued under clause 5·3·3; and

  ·3   (where Insurance Option A applies) any variation in premium referred to in paragraph A·5·1 of Schedule 3.

·2   There shall be deducted from the Contract Sum:

  ·1   all Provisional Sums and the value of all work for which an Approximate Quantity is included in the Contract Bills or in the Employer's Requirements;

  ·2   the amount of each Valuation under clause 5·6·2 or 5·8·3 of items omitted in accordance with a Variation required by the Architect/Contract Administrator or subsequently sanctioned by him in writing, together with the amount included in the Contract Bills or in the CDP Analysis for any such other work as is referred to in clause 5·9;

  ·3   any amounts deductible under clause 2·10, 2·38, 3·11, 3·18·2 or 6·15·2 and any amounts allowable to the Employer under whichever Fluctuations Option applies; and

  ·4   any other amount which is required by this Contract to be deducted from the Contract Sum.

·3   There shall be added to the Contract Sum:

  ·1   any amounts payable by the Employer to the Contractor as a result of payments made or costs incurred by the Contractor under clauses 2·21, 2·23, 3·17 and 6·5;

  ·2   the amount of the Valuation under section 5 of any Variation, including the valuation of other work as referred to in clause 5·9 but excluding any omission;

  ·3   the amount of the Valuation of work executed by, or the amount of any disbursements by, the Contractor in accordance with instructions of the Architect/Contract Administrator as to the expenditure of Provisional Sums included in the Contract Bills or in the Employer's Requirements and of all work for which an Approximate Quantity is included in the Contract Bills or in the Employer's Requirements;

continued 4·3·3

  ·4   any amounts ascertained under clause 3·24 or 4·23;

  ·5   any amounts paid by the Contractor under Insurance Option B or C or under clause 2·6·2 which the Contractor is entitled to have added to the Contract Sum;

  ·6   any amounts payable to the Contractor under whichever Fluctuations Option applies; and

  ·7   any other amount which is required by this Contract to be added to the Contract Sum.

Taking adjustments into account

4·4   Where these Conditions provide that an amount is to be added to, deducted from or dealt with by adjustment of the Contract Sum, then, as soon as the amount is ascertained in whole or in part, the ascertained amount shall be taken into account in the next computation of an Interim Certificate.

Final adjustment

4·5 ·1   Not later than 6 months after the issue by the Architect/Contract Administrator of the Practical Completion Certificate or the last Section Completion Certificate, the Contractor shall provide the Architect/Contract Administrator or (if so instructed) the Quantity Surveyor, with all documents necessary for the purposes of the adjustment of the Contract Sum.

  ·2   Not later than 3 months after receipt by the Architect/Contract Administrator or by the Quantity Surveyor of the documents referred to in clause 4·5·1:

    ·1   the Architect/Contract Administrator, or, if the Architect/Contract Administrator has so instructed, the Quantity Surveyor, shall (unless previously ascertained) ascertain the amount of any loss and/or expense under clause 3·24 or 4·23; and

    ·2   the Quantity Surveyor shall prepare a statement of all adjustments to be made to the Contract Sum pursuant to clause 4·3, other than the amount of any loss and/or expense then being ascertained under clause 4·5·2·1,

    and the Architect/Contract Administrator shall forthwith send to the Contractor a copy of that statement and (if applicable) that ascertainment.

Certificates and Payments

VAT

4·6 ·1   The Contract Sum is exclusive of VAT and in relation to any payment to the Contractor under this Contract, the Employer shall in addition pay the amount of any VAT properly chargeable in respect of it.

  ·2   To the extent that after the Base Date the supply of goods and services to the Employer becomes exempt from VAT there shall be paid to the Contractor an amount equal to the amount of input tax on the supply to the Contractor of goods and services which contribute to the Works but which as a consequence of that exemption the Contractor cannot recover.

Construction Industry Scheme (CIS)

4·7   If the Employer is or at any time up to the payment of the Final Certificate becomes a 'contractor' for the purposes of the CIS[42], the obligation of the Employer to make any payment under this Contract is subject to the provisions of the CIS.

Advance payment

4·8   Where the Contract Particulars state that clause 4·8 applies, the advance payment shall be paid to the Contractor on the date and reimbursed to the Employer on the terms stated in the Contract Particulars.

---

[42]   See the Contract Particulars (Fourth Recital and clause 4·7).

continued 4·8 Provided that where the Contract Particulars state that an advance payment bond is required, such payment shall only be made if the Contractor has provided to the Employer a bond in the terms set out in Part 1 of Schedule 6 from a surety approved by the Employer.[43]

Issue of Interim Certificates

4·9 ·1 The Architect/Contract Administrator shall from time to time as provided in clause 4·9·2 issue Interim Certificates stating the amount due to the Contractor from the Employer and specifying to what the amount relates and the basis on which that amount was calculated.

·2 Interim Certificates shall be issued on the dates provided for in the Contract Particulars up to the date of practical completion of the Works or the date within one month thereafter. Interim Certificates shall thereafter be issued as and when further amounts are ascertained as payable to the Contractor by the Employer and upon whichever is the later of the expiry of the Rectification Period or the issue of the Certificate of Making Good (or, where there are Sections, the last such period or certificate), provided always that the Architect/Contract Administrator shall not be required to issue an Interim Certificate within one calendar month of a previous Interim Certificate.

Amounts due in Interim Certificates

4·10 Subject to any agreement between the Parties as to stage payments, the amount stated as due in an Interim Certificate shall be an amount equal to the Gross Valuation pursuant to clause 4·16 less the aggregate of:

·1 any amount which may be deducted and retained by the Employer as provided in clauses 4·18 to 4·20 ('the Retention');

·2 the total amount of any advance payment due for reimbursement to the Employer in accordance with the terms for reimbursement stated in the Contract Particulars pursuant to clause 4·8; and

·3 the total amount stated as due in Interim Certificates previously issued under these Conditions.

Interim valuations

4·11 Interim valuations shall be made by the Quantity Surveyor whenever the Architect/Contract Administrator considers them necessary for the purpose of ascertaining the amount to be stated as due in an Interim Certificate, except where Fluctuations Option C (*formula adjustment*) applies[44], in which case an interim valuation shall be made before the issue of each Interim Certificate.

Application by Contractor

4·12 Without affecting the Architect/Contract Administrator's obligation to issue Interim Certificates, the Contractor, not later than 7 days before the date for issue of an Interim Certificate, may submit to the Quantity Surveyor an application setting out what the Contractor considers to be the amount of the Gross Valuation. If the Contractor submits such an application the Quantity Surveyor shall make an interim valuation. If the Quantity Surveyor disagrees with the amount shown in the Contractor's application, he at the time of making the valuation shall submit to the Contractor a statement, which shall be in similar detail to that given in the application and shall identify the disagreement.

Interim Certificates – payment

4·13 ·1 The final date for payment pursuant to an Interim Certificate shall be 14 days from the date of issue of that Interim Certificate.

·2 Notwithstanding the fiduciary interest of the Employer in the Retention as stated in clause 4·18, the Employer is entitled to exercise any right under this Contract of withholding and/or deduction from monies due or to become due to the Contractor against any amount due under an Interim Certificate, whether or not any Retention is included in that Interim Certificate by the operation of clause 4·20.

---

[43] As to approval of sureties, see the Guide.
[44] See the Contract Particulars under the reference to clause 4·21 and Schedule 7.

continued 4·13

·3 Not later than 5 days after the date of issue of an Interim Certificate the Employer shall give a written notice to the Contractor which shall, in respect of the amount stated as due in that Interim Certificate, specify the amount of the payment proposed to be made, to what the amount of the payment relates and the basis on which that amount was calculated.

·4 Not later than 5 days before the final date for payment the Employer may give a written notice to the Contractor which shall specify any amount proposed to be withheld and/or deducted from the amount due, the ground or grounds for such withholding and/or deduction and the amount of withholding and/or deduction attributable to each ground.

·5 Subject to any notice given under clause 4·13·4, the Employer shall no later than the final date for payment pay the Contractor the amount specified in the notice given under clause 4·13·3 or, in the absence of a notice under clause 4·13·3, the amount stated as due in the Interim Certificate.

·6 If the Employer fails properly to pay the amount, or any part of it, due to the Contractor under these Conditions by the final date for its payment, the Employer shall pay to the Contractor in addition to the amount not properly paid simple interest thereon at the Interest Rate for the period until such payment is made. Payment of such interest shall be treated as a debt due to the Contractor by the Employer. The acceptance of any payment of interest under this clause 4·13·6 shall not in any circumstances be construed as a waiver by the Contractor of his right to proper payment of the principal amount due from the Employer to the Contractor in accordance with these Conditions or of the rights of the Contractor to suspend performance of his obligations under this Contract pursuant to clause 4·14 or to terminate his employment under section 8.

Contractor's right of suspension

4·14 Without affecting any other rights and remedies of the Contractor, if the Employer, subject to any notice issued pursuant to clause 4·13·4, fails to pay the Contractor in full (including any VAT properly chargeable in respect of such payment) by the final date for payment as required by these Conditions and such failure continues for 7 days after the Contractor has given to the Employer, with a copy to the Architect/Contract Administrator, written notice of his intention to suspend the performance of his obligations under this Contract and the ground or grounds on which it is intended to suspend performance, then the Contractor may suspend such performance until payment in full occurs.

Final Certificate – issue and payment[45]

4·15 ·1 The Architect/Contract Administrator shall issue the Final Certificate not later than 2 months after whichever of the following occurs last:

·1 the end of the Rectification Period in respect of the Works or (where there are Sections) the last such period to expire;

·2 the date of issue of the Certificate of Making Good under clause 2·39 or (where there are Sections) the last such certificate to be issued; or

·3 the date on which the Architect/Contract Administrator sends to the Contractor copies of the statement and of any ascertainment to be prepared in compliance with clause 4·5·2.

·2 The Final Certificate shall state:

·1 the Contract Sum adjusted as necessary in accordance with clause 4·3; and

·2 the sum of the amounts already stated as due in Interim Certificates plus the amount of any advance payment paid pursuant to clause 4·8;

and the difference (if any) between the two sums shall (without affecting the rights of the Contractor in respect of any Interim Certificate which has not been paid in full by the Employer by its final date for payment) be expressed in the Final Certificate as a balance due to the Contractor from the Employer or to the Employer from the Contractor, as the case may be. The Final Certificate shall state the basis on which that amount has been calculated.

---

[45] The effect of the Final Certificate is set out in clause 1·10.

continued 4·15

·3 Not later than 5 days after the date of issue of the Final Certificate the Employer shall give a written notice to the Contractor which shall, in respect of any balance stated as due to the Contractor from the Employer in the Final Certificate, specify the amount of the payment proposed to be made, to what the amount of the payment relates and the basis on which that amount was calculated.

·4 The final date for payment of the balance payable by the Employer to the Contractor or by the Contractor to the Employer, as the case may be, shall be 28 days from the date of issue of the Final Certificate. Not later than 5 days before the final date for payment of the balance the Employer may give a written notice to the Contractor which shall specify any amount proposed to be withheld and/or deducted from any balance due to the Contractor, the ground or grounds for such withholding and/or deduction and the amount of withholding and/or deduction attributable to each ground.

·5 Where the Employer does not give a written notice pursuant to clause 4·15·3 the Employer shall, subject to any notice given under clause 4·15·4, pay the Contractor any balance stated as due to the Contractor in the Final Certificate.

·6 If the Employer or the Contractor fails properly to pay the said balance, or any part of it, by the final date for its payment the Employer or the Contractor, as the case may be, shall pay to the other in addition to the balance not properly paid simple interest thereon at the Interest Rate for the period until such payment is made. The acceptance of any payment of interest under this clause 4·15·6 shall not in any circumstances be construed as a waiver by the Contractor or the Employer of his right to proper payment of the balance.

·7 The balance due pursuant to clause 4·15·4 and any interest under clause 4·15·6 shall be treated as a debt due to the Contractor by the Employer or to the Employer by the Contractor, as the case may be.

Gross Valuation

Ascertainment

4·16 The Gross Valuation shall be the total of the amounts referred to in clauses 4·16·1 and 4·16·2 less the total of the amounts referred to in clause 4·16·3, applied up to and including a date not more than 7 days before the date of the Interim Certificate.

·1 There shall be included the following which are subject to Retention:

·1 the total value of the work properly executed by the Contractor (including any work so executed for which a value has been agreed pursuant to clause 5·2·1 or which has been valued under the Valuation Rules and work for which a Schedule 2 Quotation has been accepted), together, where applicable, with any adjustment of that value under Fluctuations Option C, but excluding any restoration, replacement or repair of loss or damage and removal and disposal of debris which under paragraphs B·3·5 and C·4·5·2 of Schedule 3 are treated as a Variation. Where there is an Activity Schedule, the value to be included in respect of the work in each activity to which it relates shall be a proportion of the price stated for the work in that activity equal to the proportion of the work in that activity that has then been properly executed;

·2 the total value of the materials and goods delivered to or adjacent to the Works for incorporation therein by the Contractor but not so incorporated, provided that the value of such materials and goods shall only be included as and from the times that they are reasonably, properly and not prematurely so delivered and have been adequately protected against weather and other casualties; and

·3 the total value of any Listed Items, the value of which is required under clause 4·17 to be included in the amount stated as due.

·2 There shall be included the following which are not subject to Retention:

·1 any amounts to be included in Interim Certificates in accordance with clause 4·4 as a result of payments made or costs incurred by the Contractor under clauses 2·6·2, 2·21, 2·23, 3·17 and 6·5 and paragraph A·5·1, B·2·1·2 or C·3·1 of Schedule 3;

continued 4·16·2

  ·2   any amounts ascertained under clause 3·24 or 4·23 or in respect of any restoration, replacement or repair of loss or damage and removal and disposal of debris which under paragraphs B·3·5 and C·4·5·2 of Schedule 3 are treated as a Variation; and

  ·3   any amount payable to the Contractor under Fluctuations Option A or B, if applicable.

·3   The following shall be deducted:

  ·1   any amounts deductible under clause 2·10, 2·38, 3·11 or 3·18·2; and

  ·2   any amount allowable by the Contractor to the Employer under Fluctuations Option A or B, if applicable.

Off-site materials and goods

4·17   The amount stated as due in an Interim Certificate shall include the value of any Listed Items before their delivery to or adjacent to the Works provided that the following conditions have been fulfilled:

·1   the Listed Items are in accordance with this Contract;

·2   the Contractor has provided the Architect/Contract Administrator with reasonable proof that:

  ·1   the property in such Listed Items is vested in the Contractor so that under clause 2·25, after the amount in respect thereof included in an Interim Certificate as properly due to the Contractor has been paid by the Employer, such Listed Items shall become the property of the Employer; and

  ·2   such Listed Items are insured against loss or damage for their full value under a policy of insurance protecting the interests of the Employer and the Contractor in respect of the Specified Perils, during the period commencing with the transfer of property in the Listed Items to the Contractor until they are delivered to, or adjacent to, the Works;

·3   at the premises where the Listed Items have been manufactured or assembled or are stored, there is in relation to such items clear identification of:

  ·1   the Employer as the person to whose order they are held; and

  ·2   their destination as the Works,

and such items either are set apart or have been clearly and visibly marked, individually or in sets, by letters or figures or by reference to a pre-determined code; and

·4   in the case of uniquely identified Listed Items, the Contractor, if it is stated in the Contract Particulars as required, has provided from a surety approved by the Employer[43] a bond in favour of the Employer in the amount specified in the Contract Particulars and in the terms set out in Part 2 of Schedule 6; or

·5   in the case of Listed Items which are not uniquely identified, the Contractor has provided from a surety approved by the Employer[43] a bond in favour of the Employer in the amount specified in the Contract Particulars and in the terms set out in Part 2 of Schedule 6.

Retention

Rules on treatment of Retention

4·18   The Retention which the Employer may deduct and retain as referred to in clause 4·10·1 shall be subject to the following rules:

·1   the Employer's interest in the Retention is fiduciary as trustee for the Contractor (but without obligation to invest);

·2   at the date of each Interim Certificate the Architect/Contract Administrator shall prepare, or instruct the Quantity Surveyor to prepare, and shall issue to the Employer and to the Contractor a statement

continued 4·18·2

specifying the amount of the Retention that has been deducted in arriving at the amount stated as due in that Interim Certificate;

·3 except where the Employer is a Local Authority, the Employer shall, to the extent that the Employer exercises his right under clause 4·20 and if the Contractor so requests, at the date of payment under each Interim Certificate place the Retention in a separate banking account (so designated as to identify the amount as the Retention held by the Employer on trust as provided in clause 4·18·1) and certify to the Architect/Contract Administrator with a copy to the Contractor that such amount has been so placed. The Employer shall be entitled to the full beneficial interest in any interest accruing on the separate banking account and shall be under no duty to account for any such interest to the Contractor or any sub-contractor;

·4 where the Employer exercises the right to withhold and/or deduct referred to in clause 4·13·2 against any Retention, he shall inform the Contractor of the amount of that withholding and/or deduction by reference to the latest statement issued under clause 4·18·2.

Retention Bond

4·19 Where the Contract Particulars state that clause 4·19 applies, then:

·1 the provisions of clauses 4·10·1 and 4·20 permitting the deduction of the Retention shall not apply save that the Architect/Contract Administrator shall at the date of each Interim Certificate prepare, or instruct the Quantity Surveyor to prepare, a statement specifying the deduction in respect of the Retention that would have been made in arriving at the amount stated as due in such Interim Certificate had those clauses applied[46];

·2 on or before the Date of Possession the Contractor shall provide to the Employer and thereafter maintain a bond ('the Retention Bond') in favour of the Employer from a surety approved by the Employer ('the Surety')[43] in the terms set out in Part 3 of Schedule 6, incorporating in clauses 2 *(maximum aggregate sum)* and 6·3 *(expiry date)* of the bond, the sum and date stated in the Contract Particulars;

·3 if the Contractor is in breach of clause 4·19·2 by not providing or not maintaining the Retention Bond, the provisions of clauses 4·10·1 and 4·20 permitting the deduction of the Retention shall apply in respect of the next Interim Certificate issued after the date of the breach and subsequent Interim Certificates. If the Contractor subsequently provides and thereafter maintains the required bond the Employer shall, in the next Interim Certificate to be issued after such compliance, have released to the Contractor the Retention deducted during the period of the breach;

·4 if at any time the amount of the Retention that would have been deducted had the provisions of clauses 4·10·1 and 4·20 applied exceeds the aggregate sum stated in the Retention Bond, then either the Contractor shall arrange with the Surety for the aggregate sum to equate to such amount or the amount not covered by the bond may be deducted by the Employer and clauses 4·18 and 4·20 shall apply to such Retention; and

·5 where the Employer has required the Contractor to provide a Performance Bond, then, in respect of any default to which that Performance Bond refers which is also a matter for which the Employer could make a demand under the terms of the Retention Bond, the Employer shall first have recourse to the Retention Bond.

Retention – rules for ascertainment

4·20 The Retention which the Employer may deduct and retain shall be such percentage of the total amount included under clause 4·16·1 in any Interim Certificate as arises from the operation of the following rules[47]:

·1 the Retention Percentage shall be 3 per cent (or such other rate as is stated in the Contract Particulars);

·2 the Retention Percentage may be deducted from so much of the said total amount as relates to:

·1 work where the Works or (where there are Sections) the Section(s) of which it forms part have not reached practical completion; and

---

[46] This saving provision is included in view of the provisions of clauses 4·2 and 4·3 of the form of Retention Bond in Schedule 6.
[47] For the effect of clause 4·20·3, see the Guide.

·2   amounts in respect of the value of materials and goods included under clauses 4·16·1·2 and 4·16·1·3;

·3   half the Retention Percentage may be deducted from so much of the said total amount as relates to work where the Works or relevant Section(s) have reached practical completion but in respect of which a Certificate of Making Good under clause 2·39 or a certificate under clause 2·35 has not been issued.

Fluctuations

Choice of fluctuation provisions

4·21   Fluctuations shall be dealt with by the application of Schedule 7 in accordance with whichever of the following is stated in the Contract Particulars to apply:

Fluctuations Option A: contribution, levy and tax fluctuations, or
Fluctuations Option B: labour and materials cost and tax fluctuations, or
Fluctuations Option C: formula adjustment.[48]

Non-applicability to Schedule 2 Quotation

4·22   Fluctuations Options A, B and C shall not apply in respect of work for which the Architect/Contract Administrator has issued to the Contractor a Confirmed Acceptance of a Schedule 2 Quotation or in respect of a Variation to such work.

Loss and Expense

Matters materially affecting regular progress

4·23   If in the execution of this Contract the Contractor incurs or is likely to incur direct loss and/or expense for which he would not be reimbursed by a payment under any other provision in these Conditions due to a deferment of giving possession of the site or relevant part of it under clause 2·5 or because the regular progress of the Works or of any part of them has been or is likely to be materially affected by any of the Relevant Matters, the Contractor may make written application to the Architect/Contract Administrator. If the Contractor makes such application, save where these Conditions provide that there shall be no addition to the Contract Sum or otherwise exclude the operation of this clause, then, if and as soon as the Architect/ Contract Administrator is of the opinion that the regular progress has been or is likely to be materially affected as stated in the application or that direct loss and/or expense has been or is likely to be incurred due to such deferment, the Architect/Contract Administrator shall from time to time thereafter ascertain, or instruct the Quantity Surveyor to ascertain, the amount of the loss and/or expense which has been or is being incurred; provided always that the Contractor shall:

·1   make his application as soon as it has become, or should reasonably have become, apparent to him that the regular progress has been or is likely to be affected;

·2   in support of his application submit to the Architect/Contract Administrator upon request such information as should reasonably enable the Architect/Contract Administrator to form an opinion; and

·3   upon request submit to the Architect/Contract Administrator or to the Quantity Surveyor such details of the loss and/or expense as are reasonably necessary for such ascertainment.

Relevant Matters

4·24   The following are the Relevant Matters:

·1   Variations (excluding any loss and/or expense relating to a Confirmed Acceptance of a Schedule 2 Quotation but including any other matters or instructions which under these Conditions are to be treated as, or as requiring, a Variation);

---

[48]   Fluctuations Option B should be used where the Parties have agreed to allow the labour and materials cost and tax fluctuations to which paragraphs B·1 to B·3 of that Option refer. Fluctuations Option C should be used where the Parties have agreed that fluctuations should be dealt with by adjustment of the Contract Sum under the JCT Formula Rules.

·2 instructions of the Architect/Contract Administrator:

   ·1 under clause 3·15 or 3·16 (excluding an instruction for expenditure of a Provisional Sum for defined work);

   ·2 for the opening up for inspection or testing of any work, materials or goods under clause 3·17 (including making good), unless the cost is provided for in the Contract Bills or unless the inspection or test shows that the work, materials or goods are not in accordance with this Contract; or

   ·3 in relation to any discrepancy in or divergence between the Contract Drawings, the Contract Bills and/or other documents referred to in clause 2·15;

·3 suspension by the Contractor under clause 4·14 of the performance of his obligations under this Contract, provided the suspension was not frivolous or vexatious;

·4 the execution of work for which an Approximate Quantity is not a reasonably accurate forecast of the quantity of work required;

·5 any impediment, prevention or default, whether by act or omission, by the Employer, the Architect/ Contract Administrator, the Quantity Surveyor or any of the Employer's Persons, except to the extent caused or contributed to by any default, whether by act or omission, of the Contractor or of any of the Contractor's Persons.

Amounts ascertained – addition to Contract Sum

4·25 Any amounts from time to time ascertained under clause 4·23 shall be added to the Contract Sum.

Reservation of Contractor's rights and remedies

4·26 The provisions of clauses 4·23 to 4·25 are without prejudice to any other rights and remedies which the Contractor may possess.

General

Definition of Variations

5·1   The term 'Variation' means:

·1   the alteration or modification of the design, quality or quantity of the Works including:

·1   the addition, omission or substitution of any work;

·2   the alteration of the kind or standard of any of the materials or goods to be used in the Works;

·3   the removal from the site of any work executed or materials or goods brought thereon by the Contractor for the purposes of the Works other than work, materials or goods which are not in accordance with this Contract;

·2   the imposition by the Employer of any obligations or restrictions in regard to the matters set out in this clause 5·1·2 or the addition to or alteration or omission of any such obligations or restrictions so imposed or imposed by the Employer in the Contract Bills or in the Employer's Requirements in regard to:

·1   access to the site or use of any specific parts of the site;

·2   limitations of working space;

·3   limitations of working hours; or

·4   the execution or completion of the work in any specific order.[49]

Valuation of Variations and provisional sum work

5·2   ·1   Subject to clause 5·2·2, the value of:

·1   all Variations required by an instruction of the Architect/Contract Administrator or subsequently sanctioned by him in writing;

·2   all work which under these Conditions is to be treated as a Variation;

·3   all work executed by the Contractor in accordance with an instruction of the Architect/Contract Administrator as to the expenditure of Provisional Sums which are included in the Contract Bills or in the Employer's Requirements; and

·4   all work executed by the Contractor for which an Approximate Quantity has been included in the Contract Bills or in the Employer's Requirements

shall be such amount as is agreed by the Employer and the Contractor or, where not agreed, shall, unless otherwise agreed by the Employer and the Contractor, be valued by the Quantity Surveyor (a 'Valuation') in accordance with clauses 5·6 to 5·10 ('the Valuation Rules'), such Valuation insofar as it relates to the Contractor's Designed Portion being in accordance with clause 5·8.

·2   Clause 5·2·1 shall not apply in respect of a Variation for which a Schedule 2 Quotation is made and for which the Architect/Contract Administrator issues a Confirmed Acceptance or in respect of a Variation thereto to which clause 5·3·3 applies.

---

[49]   See clauses 3·10·1 and (where applicable) 3·10·3 for the Contractor's right of reasonable objection to Variations.

Schedule 2 Quotation

5·3 ·1 If the Architect/Contract Administrator in his instruction for a Variation states that the Contractor is to provide a quotation in accordance with the provisions of Schedule 2 (a 'Schedule 2 Quotation'), the Contractor shall subject to receipt of sufficient information provide a quotation in accordance with those provisions, unless within 7 days of his receipt of that instruction (or such longer period as is either stated in the instruction or agreed between them) he notifies the Architect/Contract Administrator in writing that he disagrees with the application of that procedure to that instruction.

·2 If the Contractor within that 7 day period gives notice of such disagreement, he shall not be obliged to provide that quotation and the Variation shall not be carried out unless and until the Architect/Contract Administrator gives a further instruction that the Variation is to be carried out and is to be valued by a Valuation.

·3 Where a Schedule 2 Quotation has been made for work and a Confirmed Acceptance issued, then, if the Architect/Contract Administrator subsequently issues an instruction requiring a Variation of that work, the Quantity Surveyor shall make a valuation of that Variation on a fair and reasonable basis having regard to the content of that quotation and shall include in that valuation the direct loss and/or expense, if any, incurred by the Contractor because the regular progress of the Works or of any part of them is materially affected by compliance with the instruction.

Contractor's right to be present at measurement

5·4 Where it is necessary to measure work for the purpose of a Valuation the Quantity Surveyor shall give the Contractor an opportunity to be present at the time of such measurement and to take such notes and measurements as the Contractor may require.

Giving effect to Valuations, agreements etc.

5·5 Effect shall be given by means of addition to or deduction from the Contract Sum to each agreement by the Employer and the Contractor under clause 5·2·1, each Valuation, each Schedule 2 Quotation for which a Confirmed Acceptance is issued and each valuation under clause 5·3·3.

The Valuation Rules

Measurable Work

5·6 ·1 To the extent that a Valuation relates to the execution of additional or substituted work which can properly be valued by measurement or to the execution of work for which an Approximate Quantity is included in the Contract Bills and subject to clause 5·8 in the case of CDP Works, such work shall be measured and shall be valued in accordance with the following rules:

·1 where the additional or substituted work is of similar character to, is executed under similar conditions as, and does not significantly change the quantity of, work set out in the Contract Bills, the rates and prices for the work so set out shall determine the valuation;

·2 where the additional or substituted work is of similar character to work set out in the Contract Bills but is not executed under similar conditions thereto and/or significantly changes its quantity, the rates and prices for the work so set out shall be the basis for determining the valuation and the Valuation shall include a fair allowance for such difference in conditions and/or quantity;

·3 where the additional or substituted work is not of similar character to work set out in the Contract Bills, the work shall be valued at fair rates and prices;

·4 where the Approximate Quantity is a reasonably accurate forecast of the quantity of work required the rate or price for the Approximate Quantity shall determine the valuation; and

·5 where the Approximate Quantity is not a reasonably accurate forecast of the quantity of work required, the rate or price for that Approximate Quantity shall be the basis for determining the valuation and the Valuation shall include a fair allowance for such difference in quantity.

continued 5·6·1   Provided that clauses 5·6·1·4 and 5·6·1·5 shall apply only to the extent that the work has not been altered or modified other than in quantity.

·2   To the extent that a Valuation relates to the omission of work set out in the Contract Bills and subject to clause 5·8 in the case of CDP Works, the rates and prices for such work therein set out shall determine the valuation of the work omitted.

·3   In any valuation of work under clauses 5·6·1 and 5·6·2:

·1   measurement shall be in accordance with the same principles as those governing the preparation of the Contract Bills, as referred to in clause 2·13;

·2   allowance shall be made for any percentage or lump sum adjustments in the Contract Bills; and

·3   allowance, where appropriate, shall be made for any addition to or reduction of preliminary items of the type referred to in the Standard Method of Measurement, provided that no such allowance shall be made in respect of compliance with an Architect/Contract Administrator's instruction for the expenditure of a Provisional Sum for defined work.

Daywork

5·7   To the extent that a Valuation relates to the execution of additional or substituted work which cannot properly be valued by measurement the Valuation shall comprise:

·1   the prime cost of such work (calculated in accordance with the 'Definition of Prime Cost of Daywork carried out under a Building Contract' issued by The Royal Institution of Chartered Surveyors and the Construction Confederation as current at the Base Date) together with percentage additions to each section of the prime cost at the rates set out by the Contractor in the Contract Bills; or

·2   where the work is within the province of any specialist trade and The Royal Institution and the appropriate body representing the employers in that trade have agreed and issued a definition of prime cost of daywork[50], the prime cost of such work calculated in accordance with that definition current at the Base Date, together with percentage additions on the prime cost at the rates set out by the Contractor in the Contract Bills.

Provided that in any case vouchers specifying the time daily spent upon the work, the workmen's names, the plant and the materials employed shall be delivered for verification to the Architect/Contract Administrator or his authorised representative not later than the end of the week following that in which the work has been executed.

Contractor's Designed Portion – Valuation

5·8   Valuations relating to the Contractor's Designed Portion shall be made under this clause 5·8.

·1   Allowance shall be made in such Valuations for the addition or omission of the relevant design work.

·2   The valuation of additional or substituted work shall be consistent with the values of work of a similar character set out in the CDP Analysis, making due allowance for any change in the conditions under which work is carried out and/or any significant change in the quantity of the work so set out. Where there is no work of a similar character set out in the CDP Analysis a fair valuation shall be made.

·3   The valuation of the omission of work set out in the CDP Analysis shall be in accordance with the values therein for such work.

·4   Clauses 5·6·3·2, 5·6·3·3, 5·7 and 5·9 shall apply so far as is relevant.

---

[50]   There are three definitions to which clause 5·7·2 refers, namely that agreed between The Royal Institution and the Electrical Contractors Association, that agreed between The Royal Institution and the Electrical Contractors Association of Scotland and that agreed between The Royal Institution and the Heating and Ventilating Contractors Association.

Change of conditions for other work

5·9 If as a result of:

    ·1   compliance with any instruction requiring a Variation;

    ·2   compliance with any instruction as to the expenditure of a Provisional Sum for undefined work;

    ·3   compliance with any instruction as to the expenditure of a Provisional Sum for defined work, to the extent that the instruction for that work differs from the description given for such work in the Contract Bills; or

    ·4   the execution of work for which an Approximate Quantity is included in the Contract Bills, to the extent that the quantity is more or less than the quantity ascribed to that work in the Contract Bills,

there is a substantial change in the conditions under which any other work is executed (including CDP Works), then such other work shall be treated as if it had been the subject of an instruction requiring a Variation and shall be valued in accordance with the provisions of this section 5.

Additional provisions

5·10 ·1   To the extent that a Valuation does not relate to the execution of additional or substituted work or the omission of work or to the extent that the valuation of any work or liabilities directly associated with a Variation cannot reasonably be effected in the Valuation by the application of clauses 5·6 to 5·9, a fair valuation shall be made.

    ·2   No allowance shall be made under the Valuation Rules for any effect upon the regular progress of the Works or of any part of them or for any other direct loss and/or expense for which the Contractor would be reimbursed by payment under any other provision in these Conditions.

Injury to Persons and Property

Liability of Contractor – personal injury or death

6·1 The Contractor shall be liable for, and shall indemnify the Employer against, any expense, liability, loss, claim or proceedings whatsoever in respect of personal injury to or the death of any person arising out of or in the course of or caused by the carrying out of the Works, except to the extent that the same is due to any act or neglect of the Employer or of any of the Employer's Persons.

Liability of Contractor – injury or damage to property

6·2 The Contractor shall be liable for, and shall indemnify the Employer against, any expense, liability, loss, claim or proceedings in respect of any loss, injury or damage whatsoever to any property real or personal in so far as such loss, injury or damage arises out of or in the course of or by reason of the carrying out of the Works and to the extent that the same is due to any negligence, breach of statutory duty, omission or default of the Contractor or of any of the Contractor's Persons. This liability and indemnity is subject to clause 6·3 and, where Insurance Option C (Schedule 3, paragraph C·1) applies, excludes loss or damage to any property required to be insured thereunder caused by a Specified Peril.

Injury or damage to property – Works and Site Materials excluded

6·3 ·1 Subject to clauses 6·3·2 and 6·3·3, the reference in clause 6·2 to 'property real or personal' does not include the Works, work executed and/or Site Materials up to and including whichever is the earlier of:

·1 the date of issue of the Practical Completion Certificate; or

·2 the date of termination of the Contractor's employment.

·2 Where a Section Completion Certificate is issued in respect of a Section, that Section shall not after the date of issue of that certificate be regarded as 'the Works' or 'work executed' for the purpose of clause 6·3·1.

·3 If clause 2·33 has been operated, then, after the Relevant Date, the Relevant Part shall not be regarded as 'the Works' or 'work executed' for the purpose of clause 6·3·1.

Insurance against Personal Injury and Property Damage

Contractor's insurance of his liability

6·4 ·1 Without prejudice to his obligation to indemnify the Employer under clauses 6·1 and 6·2, the Contractor shall take out and maintain insurance in respect of claims arising out of his liability referred to in clauses 6·1 and 6·2 which:

·1 in respect of claims for personal injury to or the death of any employee of the Contractor arising out of and in the course of such person's employment, shall comply with all relevant legislation; and

·2 for all other claims to which clause 6·4·1 applies[51], shall indemnify the Employer in like manner to the Contractor (but only to the extent that the Contractor may be liable to indemnify the Employer under the terms of this Contract) and shall be in a sum not less than that stated in the Contract Particulars for any one occurrence or series of occurrences arising out of one event.[52]

---

[51] It should be noted that the cover granted under public liability policies taken out pursuant to clause 6·4·1 may not be co-extensive with the indemnity given to the Employer in clauses 6·1 and 6·2: for example, each claim may be subject to the excess in the policy and cover may not be available in respect of loss or damage due to gradual pollution.
[52] The Contractor may, if he wishes, insure for a sum greater than that stated in the Contract Particulars.

continued 6·4

·2   As and when reasonably required to do so by the Employer, the Contractor shall send to the Architect/ Contract Administrator for inspection by the Employer documentary evidence that the insurances required by clause 6·4·1 have been taken out and are being maintained, and at any time the Employer may (but shall not unreasonably or vexatiously) require that the relevant policy or policies and related premium receipts be sent to the Architect/Contract Administrator for such inspection.

·3   If the Contractor defaults in taking out or in maintaining insurance in accordance with clause 6·4·1 the Employer may himself insure against any liability or expense which he may incur as a result of such default and the amount paid or payable by him in respect of premiums therefor may be deducted from any monies due or to become due to the Contractor under this Contract or shall be recoverable from the Contractor as a debt.

Contractor's insurance of liability of Employer

6·5   ·1   If the Contract Particulars state that insurance under clause 6·5·1 may be required, the Contractor shall, if instructed by the Architect/Contract Administrator, take out a policy of insurance in the names of the Employer and the Contractor[53] for the amount of indemnity there stated in respect of any expense, liability, loss, claim or proceedings which the Employer may incur or sustain by reason of injury or damage to any property caused by collapse, subsidence, heave, vibration, weakening or removal of support or lowering of ground water arising out of or in the course of or by reason of the carrying out of the Works, excluding injury or damage:

·1   for which the Contractor is liable under clause 6·2; or

·2   which is attributable to errors or omissions in the designing of the Works; or

·3   which can reasonably be foreseen to be inevitable having regard to the nature of the work to be executed and the manner of its execution; or

·4   (if Insurance Option C applies) which it is the responsibility of the Employer to insure under paragraph C·1 of Schedule 3; or

·5   to the Works and Site Materials brought on to the site of the Contract for the purpose of its execution except where the Practical Completion Certificate has been issued or in so far as any Section is the subject of a Section Completion Certificate; or

·6   which arises from any consequence of war, invasion, act of foreign enemy, hostilities (whether war is declared or not), civil war, rebellion or revolution, insurrection or military or usurped power; or

·7   which is directly or indirectly caused by or contributed to by or arises from the Excepted Risks; or

·8   which is directly or indirectly caused by or arises out of pollution or contamination of buildings or other structures or of water or land or the atmosphere happening during the period of insurance, save that this exception shall not apply in respect of pollution or contamination caused by a sudden identifiable, unintended and unexpected incident which takes place in its entirety at a specific moment in time and place during the period of insurance (all pollution or contamination which arises out of one incident being considered for the purpose of this insurance to have occurred at the time such incident takes place); or

·9   which results in any costs or expenses being incurred by the Employer or in any other sums being payable by the Employer in respect of damages for breach of contract, except to the extent that such costs or expenses or damages would have attached in the absence of any contract.

·2   Any insurance under clause 6·5·1 shall be placed with insurers approved by the Employer, and the Contractor shall send to the Architect/Contract Administrator for deposit with the Employer the policy or policies and related premium receipts.

·3   The amounts expended by the Contractor to take out and maintain the insurance referred to in clause 6·5·1 shall be added to the Contract Sum.

---

[53]   A policy of insurance taken out for the purposes of clause 6·5 should not have an expiry date earlier than the end of the Rectification Period.

Excepted Risks

6·6   Notwithstanding clauses 6·1, 6·2 and 6·4·1, the Contractor shall not be liable either to indemnify the Employer or to insure against any personal injury to or the death of any person or any damage, loss or injury caused to the Works or Site Materials, work executed, the site, or any other property, by the effect of an Excepted Risk.

Insurance of the Works

Insurance Options

6·7   Insurance Options A, B and C are set out in Schedule 3. The Insurance Option that applies to this Contract is that stated in the Contract Particulars.[54]

Related definitions

6·8   In Schedule 3 and, so far as relevant, in the clauses of these Conditions the following phrases shall have the meanings given below:

All Risks Insurance[55]:   insurance which provides cover against any physical loss or damage to work executed and Site Materials and against the reasonable cost of the removal and disposal of debris and of any shoring and propping of the Works which results from such physical loss or damage but excluding the cost necessary to repair, replace or rectify:

   (a) property which is defective due to:

      (i)   wear and tear,

      (ii) obsolescence, or

      (iii) deterioration, rust or mildew;

   (b) any work executed or any Site Materials lost or damaged as a result of its own defect in design, plan, specification, material or workmanship or any other work executed which is lost or damaged in consequence thereof where such work relied for its support or stability on such work which was defective[56];

   (c) loss or damage caused by or arising from:

      (i)   any consequence of war, invasion, act of foreign enemy, hostilities (whether war be declared or not), civil war, rebellion, revolution, insurrection, military or usurped power, confiscation, commandeering, nationalisation or requisition or loss or destruction of or damage to any property by or under the order of any government *de jure* or *de facto* or public, municipal or local authority,

---

[54]   **Insurance Option A** is applicable to the erection of new buildings where the **Contractor** is required to take out a Joint Names Policy for All Risks Insurance for the Works and **Insurance Option B** is applicable where the **Employer** has elected to take out such Joint Names Policy. **Insurance Option C** is for use in the case of alterations of or extensions to existing structures; under it the **Employer** is required to take out a Joint Names Policy for All Risks Insurance for the Works and also a Joint Names Policy to insure the existing structures and their contents owned by him or for which he is responsible against loss or damage by the Specified Perils. Some Employers (e.g. tenants and homeowners) may not be able readily to obtain the Joint Names cover, in particular that under paragraph C·1. If so, Option C should not be stated to apply and consequential amendments may be necessary. See the Guide.

[55]   The definition of All Risks Insurance in clause 6·8 defines the risks for which insurance is required. Policies issued by insurers are not standardised and the way in which insurance for those risks is expressed varies.
Obtaining Terrorism Cover, which is necessary in order to comply with the requirements of Insurance Option A, B or C, will involve an additional premium and may in certain situations be difficult to effect. Where a difficulty arises discussion should take place between the Parties and their insurance advisers. See the Guide.

[56]   In any policy for All Risks Insurance taken out under Insurance Option A or B or paragraph C·2 of Insurance Option C, cover should not be reduced by the terms of any exclusion written in the policy beyond the terms of paragraph (b) in this definition of All Risks Insurance; thus an exclusion in terms 'This Policy excludes all loss of or damage to the property insured due to defective design, plan, specification, materials or workmanship' would not be in accordance with the terms of those Insurance Options or of that definition. Wider All Risks cover than that specified may be available to Contractors, though it is not standard.

continued 6·8

(ii) disappearance or shortage if such disappearance or shortage is only revealed when an inventory is made or is not traceable to an identifiable event, or

(iii) an Excepted Risk.

Excepted Risks:

ionising radiations or contamination by radioactivity from any nuclear fuel or from any nuclear waste from the combustion of nuclear fuel, radioactive toxic explosive or other hazardous properties of any explosive nuclear assembly or nuclear component thereof, pressure waves caused by aircraft or other aerial devices travelling at sonic or supersonic speeds.

Joint Names Policy:

a policy of insurance which includes the Employer and the Contractor as composite insured and under which the insurers have no right of recourse against any person named as an insured, or, pursuant to clause 6·9, recognised as an insured thereunder.

Specified Perils:

fire, lightning, explosion, storm, flood, escape of water from any water tank, apparatus or pipe, earthquake, aircraft and other aerial devices or articles dropped therefrom, riot and civil commotion, but excluding Excepted Risks.

Terrorism Cover:

insurance provided by a Joint Names Policy under Insurance Option A, B or C for physical loss or damage to work executed and Site Materials or to an existing structure and/or its contents caused by terrorism.[57]

Sub-contractors – Specified Perils cover under Joint Names All Risks Policies

6·9 ·1 The Contractor, where Insurance Option A applies, and the Employer, where Insurance Option B or C applies, shall ensure that the Joint Names Policy referred to in paragraph A·1, A·3, B·1 or C·2 of Schedule 3 shall either:

·1 provide for recognition of each sub-contractor as an insured under the relevant Joint Names Policy; or

·2 include a waiver by the relevant insurers of any right of subrogation which they may have against any such sub-contractor

in respect of loss or damage by the Specified Perils to the Works or relevant Section, work executed and Site Materials and that this recognition or waiver shall continue up to and including the date of issue of any certificate or other document which states that in relation to the Works, the sub-contractor's works are practically complete or, if earlier, the date of termination of the sub-contractor's employment. Where there are Sections and the sub-contractor's works relate to more than one Section, the recognition or waiver for such sub-contractor shall nevertheless cease in relation to a Section upon the issue of such certificate or other document for his work in that Section.

·2 The provisions of clause 6·9·1 shall apply also in respect of any Joint Names Policy taken out by the Employer under paragraph A·2, or by the Contractor under paragraph B·2·1·2 or C·3·1·2 of Schedule 3.

Terrorism Cover – non-availability – Employer's options

6·10 ·1 If the insurers named in the Joint Names Policy, or (where Insurance Option C applies) the insurers named in either or both such policies, notify either Party that, with effect from a specified date (the 'cessation date'), Terrorism Cover will cease and will no longer be available, the recipient shall immediately inform the other Party.

·2 The Employer, after receipt of such notification but before the cessation date, shall give notice to the Contractor in writing

---

[57] As respects this definition, the extent of Terrorism Cover and possible difficulties in complying with the requirements of Insurance Options A, B and C, see the Guide.

continued 6·10·2

either

·1 that, notwithstanding the cessation of Terrorism Cover, the Employer requires that the Works continue to be carried out

or

·2 that on the date stated in the Employer's notice (which shall be a date after the date of the insurers' notification but no later than the cessation date) the Contractor's employment under this Contract shall terminate.

·3 If the Employer gives notice of termination under clause 6·10·2·2, then upon and from such termination the provisions of clauses 8·12·2 to 8·12·5 (excluding clause 8·12·3·5) shall apply and the other provisions of this Contract which require any further payment or any release of Retention to the Contractor shall cease to apply.

·4 If the Employer does not give notice of termination under clause 6·10·2·2, then:

·1 if work executed and/or Site Materials suffer physical loss or damage caused by terrorism, the Contractor shall with due diligence restore the damaged work, replace or repair any lost or damaged Site Materials, remove and dispose of any debris and proceed with the carrying out of the Works;

·2 the restoration, replacement or repair of such loss or damage and (when required) the removal and disposal of debris shall be treated as a Variation, with no reduction in any amount payable to the Contractor pursuant to this clause 6·10·4 by reason of any act or neglect of the Contractor or of any sub-contractor which may have contributed to the physical loss or damage; and

·3 (where Insurance Option C applies) the requirement that the Works continue to be carried out shall not be affected by any loss or damage to the existing structures and/or their contents caused by terrorism but not so as thereby to impose any obligation on the Employer to reinstate the existing structures.

CDP Professional Indemnity Insurance

Obligation to insure

6·11 Where there is a Contractor's Designed Portion, the Contractor shall:

·1 forthwith after this Contract has been entered into, take out (unless he has already done so) a Professional Indemnity insurance policy with a limit of indemnity of the type and in an amount not less than that stated in the Contract Particulars[58];

·2 provided it remains available at commercially reasonable rates, maintain such insurance until the expiry of the period stated in the Contract Particulars from the date of practical completion of the Works; and

·3 as and when reasonably requested to do so by the Employer or the Architect/Contract Administrator, produce for inspection documentary evidence that such insurance has been effected and/or is being maintained.

Increased cost and non-availability

6·12 If the insurance referred to in clause 6·11 ceases to be available at commercially reasonable rates, the Contractor shall immediately give notice to the Employer so that the Contractor and the Employer can discuss the means of best protecting the respective positions of the Employer and the Contractor in the absence of such insurance.

---

[58] See the Guide.

Joint Fire Code – compliance

Application of clauses

6·13   Clauses 6·14 to 6·16 apply where the Contract Particulars state that the Joint Fire Code applies.

Compliance with Joint Fire Code

6·14   The Parties shall comply with the Joint Fire Code; the Employer shall ensure such compliance by all Employer's Persons and the Contractor shall ensure such compliance by all Contractor's Persons.

Breach of Joint Fire Code – Remedial Measures

6·15   ·1   If a breach of the Joint Fire Code occurs and the insurers under the Joint Names Policy in respect of the Works specify by notice to the Employer or the Contractor the remedial measures they require (the 'Remedial Measures'), the Party receiving the notice shall send copies of it to the other and to the Architect/Contract Administrator, and then:

   ·1   subject to clause 6·15·1·2, where the Remedial Measures relate to the obligation of the Contractor to carry out and complete the Works, the Contractor shall ensure that the Remedial Measures are carried out by such date as the insurers specify; and

   ·2   to the extent that the Remedial Measures require a Variation to the Works as described in the Contract Documents or in an Architect/Contract Administrator's instruction, the Architect/Contract Administrator shall issue such instructions as are necessary to enable compliance. If, in any emergency, compliance with the Remedial Measures in whole or in part requires the Contractor to supply materials or execute work before receiving instructions under this clause 6·15·1·2, the Contractor shall supply such limited materials and execute such limited work as are reasonably necessary to secure immediate compliance. The Contractor shall forthwith inform the Architect/Contract Administrator of the emergency and of the steps he is taking under this clause 6·15·1·2. Save to the extent they relate to the Contractor's Designed Portion, such work executed and materials supplied by the Contractor shall be treated as if they had been executed and supplied under an instruction requiring a Variation.

   ·2   If the Contractor, within 7 days of receipt of a notice specifying Remedial Measures not requiring an Architect/Contract Administrator's instruction under clause 6·15·1·2, does not begin to carry out or thereafter fails without reasonable cause regularly and diligently to proceed with the Remedial Measures, then the Employer may employ and pay other persons to carry out those Remedial Measures. The Contractor shall be liable for all additional costs incurred by the Employer in connection with such employment and an appropriate deduction shall be made from the Contract Sum.

Joint Fire Code – amendments/revisions

6·16   If after the Base Date the Joint Fire Code is amended or revised and the Joint Fire Code as amended or revised is, under the Joint Names Policy, applicable to the Works, the cost, if any, of compliance by the Contractor with any amendment or revision to the Joint Fire Code shall be borne as stated in the Contract Particulars. If it is to be borne by the Employer, it shall be added to the Contract Sum.

Assignment

General

7·1   Subject to clause 7·2, neither the Employer nor the Contractor shall without the written consent of the other assign this Contract or any rights thereunder.

Rights of enforcement

7·2   Where clause 7·2 is stated in the Contract Particulars to apply, then, in the event of transfer by the Employer of his freehold or leasehold interest in, or of a grant by the Employer of a leasehold interest in, the whole of the premises comprising the Works or (if the Contract Particulars so state) any Section, the Employer may at any time after practical completion of the Works or of the relevant Section grant or assign to any such transferee or lessee the right to bring proceedings in the name of the Employer (whether by arbitration or litigation, whichever applies under this Contract) to enforce any of the terms of this Contract made for the benefit of the Employer. The assignee shall be estopped from disputing any enforceable agreements reached between the Employer and the Contractor which arise out of and relate to this Contract (whether or not they are or appear to be a derogation from the right assigned) and which are made prior to the date of any grant or assignment.

Clauses 7A to 7E – Preliminary

References

7·3   In these Conditions references to Part 2 of the Contract Particulars include any document referred to there.

Notices

7·4   Each notice referred to in clauses 7A to 7E shall be in writing and given to the Contractor by actual, special or recorded delivery, with respect to which clause 8·2·3 shall apply. In the case of any collateral warranty other than a specified JCT Collateral Warranty a copy of the warranty shall accompany the notice.

Execution of Collateral Warranties

7·5   Where this Contract is executed as a deed, any collateral warranty to be entered into or procured pursuant to this section 7 shall be executed as a deed. Where this Contract is executed under hand, any such warranty may be executed under hand.

Third Party Rights from Contractor

7A   **Rights for Purchasers and Tenants**

7A·1   Where clause 7A is stated in Part 2 of the Contract Particulars to apply to a Purchaser or Tenant, P&T Rights shall vest in that Purchaser or Tenant on the date of receipt by the Contractor of the Employer's notice to that effect, stating the name of the Purchaser or Tenant and the nature of his interest in the Works.

7A·2   The rights of the Employer and/or the Contractor:

     ·1   to terminate the Contractor's employment under this Contract (whether under section 8 or otherwise), or to agree to rescind this Contract;

     ·2   to agree to amend or otherwise vary or to waive any terms of this Contract;

     ·3   to agree to settle any dispute or other matter arising out of or in connection with this Contract, in each case in or on such terms as they shall in their absolute discretion think fit, *continued 7A·2* shall not be subject to the consent of any Purchaser or Tenant.

7A·3 Where P&T Rights have vested in any Purchaser or Tenant, then, notwithstanding the provisions of clause 7A·2, the Employer and the Contractor shall not be entitled without the consent of such Purchaser or Tenant to amend or vary the express provisions of this clause 7A or of Part 1 of Schedule 5 (Third Party Rights for Purchasers and Tenants).

7B   Rights for a Funder

7B·1 Where clause 7B is stated in Part 2 of the Contract Particulars to apply to a Funder, the Employer may by notice to the Contractor confer Funder Rights on the Funder identified in the notice. Those rights shall vest in the Funder on the date of receipt by the Contractor of the Employer's notice.

7B·2 Where Funder Rights have been vested in the Funder pursuant to clause 7B·1:

·1   no amendment or variation shall be made to the express terms of this clause 7B or of Part 2 of Schedule 5 (Third Party Rights for a Funder) without the prior written consent of the Funder; and

·2   neither the Employer nor the Contractor shall agree to rescind this Contract, and the rights of the Contractor to terminate his employment under this Contract or to treat it as repudiated shall in all respects be subject to the provisions of paragraph 6 of Part 2 of Schedule 5

but, subject thereto, unless and until the Funder gives notice under paragraph 5 or paragraph 6·4 of Part 2 of Schedule 5, the Contractor shall remain free without the consent of the Funder to agree with the Employer to amend or otherwise vary or to waive any term of this Contract and to settle any dispute or other matter arising out of or in connection with this Contract, in each case in such terms as they think fit, without any requirement that the Contractor obtain the consent of the Funder.

Collateral Warranties

Contractor's Warranties – Purchasers and Tenants

7C Where clause 7C is stated in Part 2 of the Contract Particulars to apply to a Purchaser or Tenant, the Employer may by notice to the Contractor, identifying the Purchaser or Tenant and his interest in the Works, require that the Contractor within 14 days from receipt of that notice enter into with such Purchaser or Tenant a Collateral Warranty in the form CWa/P&T, completed in accordance with the P&T Rights Particulars.

Contractor's Warranty – Funder

7D Where clause 7D is stated in Part 2 of the Contract Particulars to apply to a Funder, the Employer may by notice to the Contractor require that the Contractor within 14 days from receipt of the Employer's notice enter into a Collateral Warranty with the Funder in the form CWa/F, completed in accordance with the Funder Rights Particulars.

Sub-Contractors' Warranties

7E Where Part 2 of the Contract Particulars provides for the giving by any sub-contractor of a Collateral Warranty to a Purchaser, Tenant or Funder or to the Employer, the Contractor shall within 21 days from receipt of the Employer's notice, identifying the relevant sub-contractor, type of warranty and beneficiary, comply with the Contract Documents as to obtaining such warranties in the form SCWa/P&T, SCWa/F or SCWa/E (as the case may be), completed in accordance with Part 2 of the Contract Particulars and subject to any amendments proposed by any such sub-contractor and approved by the Contractor and the Employer, such approval not to be unreasonably delayed or withheld.

General

Meaning of insolvency

8·1 For the purposes of these Conditions, a Party is Insolvent if:

·1 he enters into an arrangement, compromise or composition in satisfaction of his debts (excluding a scheme of arrangement as a solvent company for the purposes of amalgamation or reconstruction); or

·2 without a declaration of solvency, he passes a resolution or makes a determination that he be wound up; or

·3 he has a winding up order or bankruptcy order made against him; or

·4 he has appointed to him an administrator or administrative receiver; or

·5 he is the subject of any analogous arrangement, event or proceedings in any other jurisdiction; or

·6 (additionally, in the case of a partnership) each partner is the subject of an individual arrangement or any other event or proceedings referred to in clauses 8·1·1 to 8·1·5.

Notices under section 8

8·2 ·1 Notice of termination of the Contractor's employment shall not be given unreasonably or vexatiously.

·2 Such termination shall take effect on receipt of the relevant notice.

·3 Each notice referred to in this section shall be in writing and given by actual, special or recorded delivery. Where given by special or recorded delivery it shall, subject to proof to the contrary, be deemed to have been received on the second Business Day after the date of posting.

Other rights, reinstatement

8·3 ·1 The provisions of clauses 8·4 to 8·7 are without prejudice to any other rights and remedies of the Employer. The provisions of clauses 8·9 and 8·10 and (in the case of termination under either of those clauses) the provisions of clause 8·12, are without prejudice to any other rights and remedies of the Contractor.

·2 Irrespective of the grounds of termination, the Contractor's employment may at any time be reinstated if and on such terms as the Parties may agree.

Termination by Employer

Default by Contractor

8·4 ·1 If, before practical completion of the Works, the Contractor:

·1 without reasonable cause wholly or substantially suspends the carrying out of the Works or the design of the Contractor's Designed Portion; or

·2 fails to proceed regularly and diligently with the Works or the design of the Contractor's Designed Portion; or

continued 8·4·1

  ·3   refuses or neglects to comply with a written notice or instruction from the Architect/Contract Administrator requiring him to remove any work, materials or goods not in accordance with this Contract and by such refusal or neglect the Works are materially affected; or

  ·4   fails to comply with clause 3·7 or 7·1; or

  ·5   fails to comply with clause 3·25,

the Architect/Contract Administrator may give to the Contractor a notice specifying the default or defaults (the 'specified default or defaults').

·2   If the Contractor continues a specified default for 14 days from receipt of the notice under clause 8·4·1, the Employer may on, or within 10 days from, the expiry of that 14 day period by a further notice to the Contractor terminate the Contractor's employment under this Contract.

·3   If the Employer does not give the further notice referred to in clause 8·4·2, (whether as a result of the ending of any specified default or otherwise) but the Contractor repeats a specified default (whether previously repeated or not) then, upon or within a reasonable time after such repetition, the Employer may by notice to the Contractor terminate that employment.

Insolvency of Contractor

8·5 ·1   If the Contractor is Insolvent, the Employer may at any time by notice to the Contractor terminate the Contractor's employment under this Contract.

·2   The Contractor shall immediately inform the Employer in writing if he makes any proposal, gives notice of any meeting or becomes the subject of any proceedings or appointment relating to any of the matters referred to in clause 8·1.

·3   As from the date the Contractor becomes Insolvent, whether or not the Employer has given such notice of termination:

  ·1   the provisions of clauses 8·7·4, 8·7·5 and 8·8 shall apply as if such notice had been given and the other provisions of this Contract which require any further payment or any release of Retention shall cease to apply;

  ·2   the Contractor's obligations under Article 1 and these Conditions to carry out and complete the Works and the design of the Contractor's Designed Portion shall be suspended; and

  ·3   the Employer may take reasonable measures to ensure that the site, the Works and Site Materials are adequately protected and that such Site Materials are retained on site; the Contractor shall allow and shall not hinder or delay the taking of those measures.

Corruption

8·6 The Employer shall be entitled by notice to the Contractor to terminate the Contractor's employment under this or any other contract with the Employer if, in relation to this or any other such contract, the Contractor or any person employed by him or acting on his behalf shall have committed an offence under the Prevention of Corruption Acts 1889 to 1916, or, where the Employer is a Local Authority, shall have given any fee or reward the receipt of which is an offence under sub-section (2) of section 117 of the Local Government Act 1972.

Consequences of termination under clauses 8·4 to 8·6

8·7 If the Contractor's employment is terminated under clause 8·4, 8·5 or 8·6:

·1   the Employer may employ and pay other persons to carry out and complete the Works and/or (where applicable) the design for the Contractor's Designed Portion and to make good any defects of the kind referred to in clause 2·38, and he and they may enter upon and take possession of the site and the Works and (subject to obtaining any necessary third party consents) may use all temporary buildings, plant, tools, equipment and Site Materials for those purposes;

   ·2   the Contractor shall:

      ·1   when required in writing by the Architect/Contract Administrator to do so (but not before), remove or procure the removal from the Works of any temporary buildings, plant, tools, equipment, goods and materials belonging to the Contractor or Contractor's Persons;

      ·2   (where there is a Contractor's Designed Portion) without charge provide the Employer with 2 copies of all Contractor's Design Documents then prepared, whether or not previously provided;

      ·3   if so required by the Employer (or by the Architect/Contract Administrator on his behalf) within 14 days of the date of termination, assign (so far as assignable and so far as he may lawfully be required to do so) to the Employer, without charge, the benefit of any agreement for the supply of materials or goods and/or for the execution of any work for the purposes of this Contract[59];

   ·3   (if not already applicable) clauses 8·7·4, 8·7·5 and 8·8 shall thereupon apply and the other provisions of this Contract which require any further payment or any release of Retention to the Contractor shall cease to apply;

   ·4   within a reasonable time after the completion of the Works and the making good of defects (or of instructions otherwise, as referred to in clause 2·38), an account of the following shall be set out in a certificate issued by the Architect/Contract Administrator or a statement prepared by the Employer:

      ·1   the amount of expenses properly incurred by the Employer, including those incurred pursuant to clause 8·7·1 and, where applicable, clause 8·5·3·3, and of any direct loss and/or damage caused to the Employer and for which the Contractor is liable, whether arising as a result of the termination or otherwise;

      ·2   the amount of payments made to the Contractor; and

      ·3   the total amount which would have been payable for the Works in accordance with this Contract;

   ·5   if the sum of the amounts stated under clauses 8·7·4·1 and 8·7·4·2 exceeds the amount stated under clause 8·7·4·3, the difference shall be a debt payable by the Contractor to the Employer or, if that sum is less, by the Employer to the Contractor.

Employer's decision not to complete the Works

8·8   ·1   If within the period of 6 months from the date of termination of the Contractor's employment the Employer decides not to have the Works carried out and completed, he shall forthwith notify the Contractor in writing. Within a reasonable time from the date of such notification, or if no notification is given but within that 6 month period the Employer does not commence to make arrangements for such carrying out and completion, then upon the expiry of that 6 month period, the Employer shall send to the Contractor a statement setting out:

      ·1   the total value of work properly executed at the date of termination or date on which the Contractor became Insolvent, ascertained in accordance with these Conditions as if that employment had not been terminated, together with any amounts due to the Contractor under these Conditions not included in such total value; and

      ·2   the aggregate amount of any expenses properly incurred by the Employer and of any direct loss and/or damage caused to the Employer and for which the Contractor is liable, whether arising as a result of the termination or otherwise.

   ·2   After taking into account amounts previously paid to the Contractor under this Contract, if the amount stated under clause 8·8·1·2 exceeds the amount stated under clause 8·8·1·1, the difference shall be a debt payable by the Contractor to the Employer or, if the clause 8·8·1·2 amount is less, by the Employer to the Contractor.

---

[59]   Clause 8·7·2·3 may not be effectual in cases of Contractor's insolvency.

Termination by Contractor

Default by Employer

8·9 ·1 If the Employer:

    ·1 does not pay by the final date for payment the amount properly due to the Contractor in respect of any certificate and/or any VAT properly chargeable on that amount; or

    ·2 interferes with or obstructs the issue of any certificate due under this Contract; or

    ·3 fails to comply with clause 7·1; or

    ·4 fails to comply with clause 3·25,

    the Contractor may give to the Employer a notice specifying the default or defaults (the 'specified default or defaults').

·2 If before practical completion of the Works the carrying out of the whole or substantially the whole of the uncompleted Works is suspended for a continuous period of the length stated in the Contract Particulars by reason of:

    ·1 Architect/Contract Administrator's instructions under clause 2·15, 3·14 or 3·15; and/or

    ·2 any impediment, prevention or default, whether by act or omission, by the Employer, the Architect/ Contract Administrator, the Quantity Surveyor or any of the Employer's Persons

    (but in either case excluding such instructions as are referred to in clause 8·11·1·2), then, unless in either case that is caused by the negligence or default of the Contractor or of any of the Contractor's Persons, the Contractor may give to the Employer a notice specifying the event or events (the 'specified suspension event or events').

·3 If a specified default or a specified suspension event continues for 14 days from the receipt of notice under clause 8·9·1 or 8·9·2, the Contractor may on, or within 10 days from, the expiry of that 14 day period by a further notice to the Employer terminate the Contractor's employment under this Contract.

·4 If the Contractor for any reason does not give the further notice referred to in clause 8·9·3, but (whether previously repeated or not):

    ·1 the Employer repeats a specified default; or

    ·2 a specified suspension event is repeated for any period, such that the regular progress of the Works is or is likely to be materially affected thereby,

    then, upon or within a reasonable time after such repetition, the Contractor may by notice to the Employer terminate the Contractor's employment under this Contract.

Insolvency of Employer

8·10 ·1 If the Employer is Insolvent, the Contractor may by notice to the Employer terminate the Contractor's employment under this Contract;

·2 the Employer shall immediately inform the Contractor in writing if he makes any proposal, gives notice of any meeting or becomes the subject of any proceedings or appointment relating to any of the matters referred to in clause 8·1;

·3 as from the date the Employer becomes Insolvent, the Contractor's obligations under Article 1 and these Conditions to carry out and complete the Works and the design of the Contractor's Designed Portion shall be suspended.

Termination by either Party

8·11 ·1 If, before practical completion of the Works, the carrying out of the whole or substantially the whole of the uncompleted Works is suspended for the relevant continuous period of the length stated in the Contract Particulars by reason of one or more of the following events:

  ·1 force majeure;

  ·2 Architect/Contract Administrator's instructions under clause 2·15, 3·14 or 3·15 issued as a result of the negligence or default of any Statutory Undertaker;

  ·3 loss or damage to the Works occasioned by any of the Specified Perils;

  ·4 civil commotion or the use or threat of terrorism and/or the activities of the relevant authorities in dealing with such event or threat; or

  ·5 the exercise by the United Kingdom Government of any statutory power which directly affects the execution of the Works,

  then either Party, subject to clause 8·11·2, may upon the expiry of that relevant period of suspension give notice in writing to the other that, unless the suspension ceases within 7 days after the date of receipt of that notice, he may terminate the Contractor's employment under this Contract. Failing such cessation within that 7 day period, he may then by further notice terminate that employment.

·2 The Contractor shall not be entitled to give notice under clause 8·11·1 in respect of the matter referred to in clause 8·11·1·3 where the loss or damage to the Works occasioned by a Specified Peril was caused by the negligence or default of the Contractor or of any of the Contractor's Persons.

Consequences of Termination under clauses 8·9 to 8·11, etc.

8·12 If the Contractor's employment is terminated under any of clauses 8·9 to 8·11, under clause 6·10·2·2 or under paragraph C·4·4 of Schedule 3:

·1 the provisions of this clause 8·12 shall thereupon apply and the other provisions of this Contract which require any further payment or any release of Retention to the Contractor shall cease to apply;

·2 the Contractor shall:

  ·1 with all reasonable dispatch remove or procure the removal from the site of any temporary buildings, plant, tools and equipment belonging to the Contractor and Contractor's Persons and, subject to the provisions of clause 8·12·5, all goods and materials (including Site Materials); and

  ·2 (where there is a Contractor's Designed Portion) without charge provide to the Employer 2 copies of the documents referred to in clause 2·40 then prepared;

·3 where the Contractor's employment is terminated under clause 8·9 or 8·10, the Contractor shall as soon as reasonably practical prepare an account or, where terminated under clause 8·11 or 6·10·2·2 or under paragraph C·4·4 of Schedule 3, the Contractor shall at the Employer's option either so prepare that account or, not later than 2 months after the date of termination, provide the Employer with all documents necessary for the Employer to prepare it, which the Employer shall do with reasonable dispatch (and in any event within 3 months of receipt of such documents). The account shall set out the amounts referred to in clauses 8·12·3·1 to 8·12·3·4 and, if applicable, clause 8·12·3·5, namely:

  ·1 the total value of work properly executed at the date of termination of the Contractor's employment, ascertained in accordance with these Conditions as if the employment had not been terminated, together with any other amounts due to the Contractor under these Conditions;

  ·2 any sums ascertained in respect of direct loss and/or expense under clauses 3·24 and 4·23 (whether ascertained before or after the date of termination);

  ·3 the reasonable cost of removal under clause 8·12·2;

continued 8·12·3

·4   the cost of materials or goods (including Site Materials) properly ordered for the Works for which the Contractor then has paid or is legally bound to pay;

·5   any direct loss and/or damage caused to the Contractor by the termination;

·4   the account shall include the amount, if any, referred to in clause 8·12·3·5 only where the Contractor's employment is terminated either:

·1   under clause 8·9 or 8·10; or

·2   under clause 8·11·1·3, if the loss or damage to the Works occasioned by any of the Specified Perils was caused by the negligence or default of the Employer or of any of the Employer's Persons;

·5   after taking into account amounts previously paid to the Contractor under this Contract, the Employer shall pay to the Contractor the amount properly due in respect of the account within 28 days of its submission by the Employer to the Contractor (or vice versa), without deduction of any Retention. Payment by the Employer for any such materials and goods as are referred to in clause 8·12·3·4 shall be subject to such materials and goods thereupon becoming the property of the Employer.

Mediation

9·1   The Parties may by agreement seek to resolve any dispute or difference arising under this Contract through mediation.[60]

Adjudication

9·2   If a dispute or difference arises under this Contract which either Party wishes to refer to adjudication, the Scheme shall apply, subject to the following:

·1   for the purposes of the Scheme the Adjudicator shall be the person (if any) and the nominating body shall be that stated in the Contract Particulars;

·2   where the dispute or difference is or includes a dispute or difference relating to clause 3·18·4 and as to whether an instruction issued thereunder is reasonable in all the circumstances:

·1   the Adjudicator to decide such dispute or difference shall (where practicable) be an individual with appropriate expertise and experience in the specialist area or discipline relevant to the instruction or issue in dispute;

·2   if the Adjudicator does not have the appropriate expertise and experience, the Adjudicator shall appoint an independent expert with such expertise and experience to advise and report in writing on whether or not the instruction under clause 3·18·4 is reasonable in all the circumstances.

Arbitration

Conduct of arbitration

9·3   Any arbitration pursuant to Article 8 shall be conducted in accordance with the JCT 2005 edition of the Construction Industry Model Arbitration Rules (CIMAR), provided that if any amendments to that edition of the Rules have been issued by the JCT the Parties may, by a joint notice in writing to the Arbitrator, state that they wish the arbitration to be conducted in accordance with the Rules as so amended. References in clause 9·4 to a Rule or Rules are references to such Rule(s) as set out in the JCT 2005 edition of CIMAR.[61]

Notice of reference to arbitration

9·4   ·1   Where pursuant to Article 8 either Party requires a dispute or difference to be referred to arbitration, that Party shall serve on the other Party a written notice of arbitration to such effect in accordance with Rule 2.1 identifying the dispute and requiring the other Party to agree to the appointment of an arbitrator. The Arbitrator shall be an individual agreed by the Parties or, failing such agreement within 14 days (or any agreed extension of that period) after the notice of arbitration is served, appointed on the application of either Party in accordance with Rule 2.3 by the person named in the Contract Particulars.

·2   Where two or more related arbitral proceedings in respect of the Works fall under separate arbitration agreements, Rules 2.6, 2.7 and 2.8 shall apply.

·3   After an arbitrator has been appointed either Party may give a further notice of arbitration to the other Party and to the Arbitrator referring any other dispute which falls under Article 8 to be decided in the arbitral proceedings and Rule 3.3 shall apply.

---

[60]   See the Guide.
[61]   Arbitration or legal proceedings are **not** an appeal against the decision of the Adjudicator but are a consideration of the dispute or difference as if no decision had been made by an Adjudicator.

Powers of Arbitrator

9·5 Subject to the provisions of Article 8 and clause 1·10 the Arbitrator shall, without prejudice to the generality of his powers, have power to rectify this Contract so that it accurately reflects the true agreement made by the Parties, to direct such measurements and/or valuations as may in his opinion be desirable in order to determine the rights of the Parties and to ascertain and award any sum which ought to have been the subject of or included in any certificate and to open up, review and revise any certificate, opinion, decision, requirement or notice and to determine all matters in dispute which shall be submitted to him in the same manner as if no such certificate, opinion, decision, requirement or notice had been given.

Effect of award

9·6 Subject to clause 9·7 the award of the Arbitrator shall be final and binding on the Parties.

Appeal – questions of law

9·7 The Parties hereby agree pursuant to section 45(2)(a) and section 69(2)(a) of the Arbitration Act 1996 that either Party may (upon notice to the other Party and to the Arbitrator):

·1 apply to the courts to determine any question of law arising in the course of the reference; and

·2 appeal to the courts on any question of law arising out of an award made in an arbitration under this arbitration agreement.

Arbitration Act 1996

9·8 The provisions of the Arbitration Act 1996 shall apply to any arbitration under this Contract wherever the same, or any part of it, shall be conducted.

Schedules

Schedule 1 Contractor's Design Submission Procedure

(Clause 2·9·3)

1 The Contractor shall prepare and submit two copies of each of the Contractor's Design Documents to the Architect/Contract Administrator in such format as is stated in the Employer's Requirements or the Contractor's Proposals and in sufficient time to allow any comments of the Architect/Contract Administrator to be incorporated prior to the relevant Contractor's Design Document being used for procurement and/or in the carrying out of the CDP Works.

2 Within 14 days from the date of receipt of any Contractor's Design Document, or (if later) 14 days from either the date or expiry of the period for submission of the same stated in the Contract Documents, the Architect/ Contract Administrator shall return one copy of that Contractor's Design Document to the Contractor marked 'A', 'B' or 'C' provided that a document shall be marked 'B' or 'C' only where the Architect/Contract Administrator considers that it is not in accordance with this Contract.

3 If the Architect/Contract Administrator does not respond to a Contractor's Design Document in the time stated in paragraph 2, it shall be regarded as marked 'A'.

4 Where the Architect/Contract Administrator marks a Contractor's Design Document 'B' or 'C', he shall identify by means of a written comment why he considers that it is not in accordance with this Contract.

5 When a Contractor's Design Document is returned by the Architect/Contract Administrator:

·1 if it is marked 'A', the Contractor shall carry out the CDP Works in strict accordance with that document;

·2 if it is marked 'B', the Contractor may carry out the CDP Works in accordance with that document, provided that the Architect/Contract Administrator's comments are incorporated into it and an amended copy of it is promptly submitted to the Architect/Contract Administrator; or

·3 if it is marked 'C', the Contractor shall take due account of the Architect/Contract Administrator's comments on it and shall either forthwith resubmit it to the Architect/Contract Administrator in amended form for comment in accordance with paragraph 1 or notify the Architect/Contract Administrator under paragraph 7.

6 The Contractor shall not carry out any work in accordance with a Contractor's Design Document marked 'C' and the Employer shall not be liable to pay for any work within the CDP Works executed otherwise than in accordance with Contractor's Design Documents marked 'A' or 'B'.

7 If the Contractor disagrees with a comment of the Architect/Contract Administrator and considers that the Contractor's Design Document in question is in accordance with this Contract, he shall within 7 days of receipt of the comment notify the Architect/Contract Administrator in writing that he considers that compliance with the comment would give rise to a Variation. Such notification shall be accompanied by a statement setting out the Contractor's reasons. Upon receipt of such a notification the Architect/Contract Administrator shall within 7 days either confirm or withdraw the comment and, where the comment is confirmed, the Contractor shall amend and resubmit the document accordingly.

8 Provided always that:

·1 confirmation or withdrawal of a comment in accordance with paragraph 7 shall not signify acceptance by either the Employer or the Architect/Contract Administrator that the relevant Contractor's Design

continued 8·1

Document or amended document is in accordance with this Contract or that compliance with the Architect/Contract Administrator's comment would give rise to a Variation;

·2 where in relation to a comment by the Architect/Contract Administrator the Contractor does not notify him in accordance with paragraph 7, the comment in question shall not be treated as giving rise to a Variation; and

·3 neither compliance with the design submission procedure in this Schedule nor with the Architect/ Contract Administrator's comments shall diminish the Contractor's obligations to ensure that the Contractor's Design Documents and CDP Works are in accordance with this Contract.

Submission of Quotation

1 ·1   Any instruction of the Architect/Contract Administrator requesting a Schedule 2 Quotation shall provide sufficient information[62] to enable the Contractor to provide that quotation, which shall comprise the matters set out in paragraph 2 of this Schedule, in compliance with the instruction. If the Contractor reasonably considers that the information provided is not sufficient, then, not later than 7 days from the receipt of the instruction, he shall notify the Architect/Contract Administrator who shall supply that information.

 ·2   The Contractor shall submit his Schedule 2 Quotation to the Quantity Surveyor in compliance with the instruction not later than 21 days from the later of:

  ·1   the date of receipt of the instruction; or

  ·2   the date of receipt by the Contractor of sufficient information as referred to in paragraph 1·1.

 ·3   The Schedule 2 Quotation shall remain open for acceptance by the Employer for 7 days from its receipt by the Quantity Surveyor.

 ·4   The Variation for which the Contractor has submitted his Schedule 2 Quotation shall not be carried out by the Contractor until receipt by the Contractor of the Confirmed Acceptance issued by the Architect/Contract Administrator under paragraph 3·2.

Content of the Quotation

2   The Schedule 2 Quotation shall separately comprise:

 ·1   the amount of the adjustment to the Contract Sum (other than any amount to which paragraph 2·3 refers), including the effect of the instruction on any other work supported by all necessary calculations, which shall be made by reference, where relevant, to the rates and prices in the Contract Bills, and including also, where appropriate, allowances for any adjustment of preliminary items;

 ·2   any adjustment to the time required for completion of the Works and/or any Section (including, where relevant, stating an earlier Completion Date than the Date for Completion given in the Contract Particulars) to the extent that such adjustment is not included in any revision of the Completion Date that has been made by the Architect/Contract Administrator under clause 2·28 or in his Confirmed Acceptance of any other Schedule 2 Quotation;

 ·3   the amount to be paid in lieu of any ascertainment under clause 4·23 of direct loss and/or expense not included in any other accepted Schedule 2 Quotation or in any previous ascertainment under clause 4·23;

 ·4   a fair and reasonable amount in respect of the cost of preparing the Schedule 2 Quotation; and

 ·5   where specifically required by the instruction, shall provide indicative information in statements on:

  ·1   the additional resources (if any) required to carry out the Variation; and

  ·2   the method of carrying out the Variation.

Each part of the Schedule 2 Quotation shall contain reasonably sufficient supporting information to enable that part to be evaluated by or on behalf of the Employer.

---

[62]   The information provided to the Contractor should normally be in a similar format to that provided at the tender stage, whether in the form of drawings and/or an addendum bill of quantities and/or a specification or otherwise. If an addendum bill is provided, see clauses 2·13 and 2·14.

Acceptance of the Quotation

3 ·1 If the Employer wishes to accept a Schedule 2 Quotation the Employer shall so notify the Contractor in writing not later than the last day of the period for acceptance stated in paragraph 1·3.

·2 If the Employer accepts a Schedule 2 Quotation the Architect/Contract Administrator shall, immediately upon that acceptance, confirm such acceptance by stating in writing to the Contractor (in this Schedule and elsewhere in the Conditions a 'Confirmed Acceptance'):

·1 that the Contractor is to carry out the Variation;

·2 the adjustment of the Contract Sum, including any amounts to which paragraphs 2·3 and 2·4 refer, to be made for complying with the instruction requiring the Variation; and

·3 any adjustment to the time required by the Contractor for completion of the Works and/or Section and the revised Completion Date(s) arising therefrom (which, where relevant, may be a date earlier than the Date for Completion given in the Contract Particulars).

Quotation not accepted

4 If the Employer does not accept the Schedule 2 Quotation by the expiry of the period for acceptance stated in paragraph 1·3, the Architect/Contract Administrator shall on the expiry of that period either:

·1 instruct that the Variation is to be carried out and is to be valued under the Valuation Rules (*clauses 5·6 to 5·10*); or

·2 instruct that the Variation is not to be carried out.

Costs of Quotation

5 If a Schedule 2 Quotation is not accepted, a fair and reasonable amount shall be added to the Contract Sum in respect of the cost of preparation of the Schedule 2 Quotation provided that the Schedule 2 Quotation has been prepared on a fair and reasonable basis. The non-acceptance by the Employer of a Schedule 2 Quotation shall not of itself be evidence that the quotation was not prepared on such a basis.

Restriction on use of Quotation

6 Unless the Architect/Contract Administrator issues a Confirmed Acceptance of a Schedule 2 Quotation, neither the Employer nor the Contractor may use that quotation for any purpose whatsoever.

Time periods

7 The Employer and the Contractor may agree to increase or reduce the number of days stated in clause 5·3·1, clause 5·3·2 or this Schedule; any such agreement shall be confirmed by the Employer to the Contractor in writing.

Insurance Option A

New Buildings – All Risks Insurance of the Works by the Contractor [63]

Contractor to take out and maintain a Joint Names Policy

A·1   The Contractor shall take out and maintain with insurers approved by the Employer a Joint Names Policy for All Risks Insurance with cover no less than that specified in clause 6·8[64] for the full reinstatement value of the Works or (where applicable) Sections (plus the percentage, if any, stated in the Contract Particulars to cover professional fees)[65] and (subject to clause 2·36) shall maintain such Joint Names Policy up to and including the date of issue of the Practical Completion Certificate or, if earlier, the date of termination of the Contractor's employment (whether or not the validity of that termination is contested).

The obligation to maintain the Joint Names Policy shall not apply in relation to a Section after the date of issue of the Section Completion Certificate for that Section.

Insurance documents – failure by Contractor to insure

A·2   The Contractor shall send to the Architect/Contract Administrator for deposit with the Employer the Joint Names Policy referred to in paragraph A·1, each premium receipt for it and any relevant endorsements of it. If the Contractor defaults in taking out or in maintaining the Joint Names Policy as required by paragraph A·1 (or fails to maintain a policy in accordance with paragraph A·3), the Employer may himself take out and maintain a Joint Names Policy against any risk in respect of which the default has occurred and the amount paid or payable by him in respect of premiums may be deducted by him from any monies due or to become due to the Contractor under this Contract or shall be recoverable from the Contractor as a debt.

Use of Contractor's annual policy – as alternative

A·3   If and so long as the Contractor independently of this Contract maintains an insurance policy which in respect of the Works or Sections:

·1   provides (inter alia) All Risks Insurance with cover and in amounts no less than those specified in paragraph A·1; and

·2   is a Joint Names Policy,

such policy shall satisfy the Contractor's obligations under paragraph A·1. The Employer may at any reasonable time inspect the policy and premium receipts for it or require that they be sent to the Architect/Contract Administrator for such inspection. So long as the Contractor, as and when reasonably required to do so, supplies the documentary evidence that the policy is being so maintained, the Contractor shall not

---

[63]   **Insurance Option A** is applicable to the erection of new buildings where the **Contractor** is required to take out a Joint Names Policy for All Risks Insurance for the Works and **Insurance Option B** is applicable where the **Employer** has elected to take out such Joint Names Policy. **Insurance Option C** is for use in the case of alterations of or extensions to existing structures; under it the **Employer** is required to take out a Joint Names Policy for All Risks Insurance for the Works and also a Joint Names Policy to insure the existing structures and their contents owned by him or for which he is responsible against loss or damage by the Specified Perils. Some Employers (e.g. tenants and homeowners) may not be able readily to obtain the Joint Names cover, in particular that under paragraph C·1. If so, Option C should not be stated to apply and consequential amendments may be necessary. See the Guide.

[64]   The definition of All Risks Insurance in clause 6·8 specifies the risks for which insurance is required. Policies issued by insurers are not standardised and the way in which the insurance for those risks is expressed varies. **In some cases it may not be possible for insurance to be taken out against certain of the risks covered by the definition of All Risks Insurance and note the potential difficulty with respect to Terrorism Cover mentioned at footnote [55].** These matters should be arranged between the Parties and their insurance advisers **prior to entering into the Contract**. See the Guide.

[65]   As to reinstatement value, irrecoverable VAT and other costs, see the Guide. As respects sub-contractors, note also the provisions of clause 6·9.

---

continued A·3 be obliged under paragraph A·2 to deposit the policy and premium receipts with the Employer. The annual renewal date of the policy, as supplied by the Contractor, is stated in the Contract Particulars.

Loss or damage, insurance claims and Contractor's obligations

A·4 ·1 If loss or damage affecting any executed work or Site Materials is occasioned by any risk covered by the Joint Names Policy, then, upon its occurrence or later discovery, the Contractor shall forthwith give notice in writing both to the Architect/Contract Administrator and to the Employer of its extent, nature and location.

·2 Subject to clause 6·10·4·2 and paragraph A·4·4, the occurrence of such loss or damage shall be disregarded in computing any amounts payable to the Contractor under this Contract.

·3 After any inspection required by the insurers in respect of a claim under the Joint Names Policy has been completed, the Contractor shall with due diligence restore the damaged work, replace or repair any lost or damaged Site Materials, remove and dispose of any debris and proceed with the carrying out and completion of the Works.

·4 The Contractor, for himself and for all his sub-contractors who pursuant to clause 6·9 are recognised as an insured under the Joint Names Policy, shall authorise the insurers to pay all monies from such insurance to the Employer. The Employer shall pay all such monies to the Contractor (less only the amount stated in paragraph A·4·5) by instalments under certificates of the Architect/Contract Administrator issued on the dates fixed for the issue of Interim Certificates.

·5 The Employer may retain from the monies paid by the insurers the amount properly incurred by the Employer in respect of professional fees up to an amount which shall not exceed the amount of the additional percentage cover for those fees or (if less) the amount paid by insurers in respect of those fees.

·6 In respect of the restoration, replacement or repair of such loss or damage and (when required) the removal and disposal of debris, the Contractor shall not be entitled to any payment other than of monies received under the Joint Names Policy.

Terrorism Cover – premium rate changes

A·5 ·1 If the rate on which the premium is based for Terrorism Cover required under the Joint Names Policy referred to in paragraph A·1 or A·3 is varied at any renewal of the cover, the Contract Sum shall be adjusted by the net amount of the difference between the premium paid by the Contractor and the premium that would have been paid but for the change in the rate.

·2 Where the Employer is a Local Authority, the Employer may, in lieu of any adjustment of the Contract Sum under paragraph A·5·1, instruct the Contractor not to renew the Terrorism Cover under the Joint Names Policy and where he so instructs, the terms of clauses 6·10·4·1 and 6·10·4·2 shall apply from the renewal date if work executed and/or Site Materials suffer physical loss or damage caused by terrorism.

Insurance Option B

New Buildings – All Risks Insurance of the Works by the Employer [63]

Employer to take out and maintain a Joint Names Policy

B·1 The Employer shall take out and maintain a Joint Names Policy for All Risks Insurance with cover no less than that specified in clause 6·8[64] for the full reinstatement value of the Works or (where applicable) Sections (plus the percentage, if any, stated in the Contract Particulars to cover professional fees)[65] and (subject to clause 2·36) shall maintain such Joint Names Policy up to and including the date of issue of the Practical Completion Certificate or, if earlier, the date of termination of the Contractor's employment (whether or not the validity of that termination is contested).

The obligation to maintain the Joint Names Policy shall not apply in relation to a Section after the date of issue of the Section Completion Certificate for that Section.

Evidence of Insurance

B·2 ·1 Except where the Employer is a Local Authority:

   ·1 the Employer shall, as and when reasonably required by the Contractor, produce documentary evidence and receipts showing that the Joint Names Policy has been taken out and is being maintained; and

   ·2 if the Employer defaults in taking out or in maintaining the Joint Names Policy, the Contractor may himself take out and maintain a Joint Names Policy against any risk in respect of which the default has occurred and the amount paid or payable by him in respect of the premiums shall be added to the Contract Sum.

   ·2 Where the Employer is a Local Authority, the Employer shall, as and when reasonably required by the Contractor, produce to the Contractor a copy of the cover certificate issued by the insurer named in the Joint Names Policy certifying that Terrorism Cover is being provided under that Policy.

Loss or damage, insurance claims, Contractor's obligations and payment by Employer

B·3 ·1 If loss or damage affecting any executed work or Site Materials is occasioned by any risk covered by the Joint Names Policy, then, upon its occurrence or later discovery, the Contractor shall forthwith give notice in writing both to the Architect/Contract Administrator and to the Employer of its extent, nature and location.

   ·2 Subject to clause 6·10·4·2 and paragraph B·3·5, the occurrence of such loss or damage shall be disregarded in computing any amounts payable to the Contractor under this Contract.

   ·3 After any inspection required by the insurers in respect of a claim under the Joint Names Policy has been completed, the Contractor shall with due diligence restore the damaged work, replace or repair any lost or damaged Site Materials, remove and dispose of any debris and proceed with the carrying out and completion of the Works.

   ·4 The Contractor, for himself and for all his sub-contractors who pursuant to clause 6·9 are recognised as an insured under the Joint Names Policy, shall authorise the insurers to pay all monies from such insurance to the Employer.

   ·5 The restoration, replacement or repair of such loss or damage and (when required) the removal and disposal of debris shall be treated as a Variation.

Insurance Option C

Insurance by the Employer of Existing Structures and Works in or Extensions to them[63]

Existing structures and contents – Joint Names Policy for Specified Perils

C·1   The Employer shall take out and maintain a Joint Names Policy in respect of the existing structures (which from the Relevant Date shall include any Relevant Part to which clause 2·33 refers) together with the contents thereof owned by him or for which he is responsible, for the full cost of reinstatement[65], repair or replacement of loss or damage due to any of the Specified Perils up to and including the date of issue of the Practical Completion Certificate or (if earlier) the date of termination of the Contractor's employment (whether or not the validity of that termination is contested). The Contractor shall authorise the insurers to pay all monies from such insurance to the Employer.

The Works – Joint Names Policy for All Risks

C·2   The Employer shall take out and maintain a Joint Names Policy for All Risks Insurance with cover no less than that specified in clause 6·8[64] for the full reinstatement value of the Works or (where applicable) Sections (plus the percentage, if any, stated in the Contract Particulars to cover professional fees)[65] and (subject to clause 2·36) shall maintain such Joint Names Policy up to and including the date of issue of the Practical Completion Certificate or, if earlier, the date of termination of the Contractor's employment (whether or not the validity of that termination is contested).

The obligation to maintain the Joint Names Policy under this paragraph C·2 shall not apply in relation to any Section after the date of issue of the Section Completion Certificate for that Section.

Evidence of Insurance

C·3   ·1   Except where the Employer is a Local Authority:

     ·1   the Employer shall, as and when reasonably required by the Contractor, produce documentary evidence and receipts showing that the Joint Names Policies required under paragraphs C·1 and C·2 have been taken out and are being maintained;

     ·2   if the Employer defaults in taking out or in maintaining either of those Joint Names Policies, the Contractor may himself take out and maintain a Joint Names Policy against any risk in respect of which the default has occurred and for that purpose, in relation to any default under paragraph C·1, shall have such right of entry and inspection as may be required to make a survey and inventory of the existing structures and the relevant contents; and

     ·3   in the event of any such default, a sum equivalent to the premiums paid or payable by the Contractor pursuant to paragraph C·3·1·2 shall be added to the Contract Sum.

   ·2   Where the Employer is a Local Authority, the Employer shall, as and when reasonably required by the Contractor, produce to the Contractor copies of the cover certificates issued by the insurers named in the Joint Names Policies under paragraphs C·1 and C·2 which certify that Terrorism Cover is being provided under each policy.

Loss or damage to Works – insurance claims and Contractor's obligations

C·4   ·1   If loss or damage affecting any executed work or Site Materials is occasioned by any of the risks covered by the Joint Names Policy referred to in paragraph C·2 then, upon its occurrence or later discovery, the Contractor shall forthwith give notice in writing both to the Architect/Contract Administrator and to the Employer of its extent, nature and location.

   ·2   Subject to clause 6·10·4·2 and paragraph C·4·5·2, the occurrence of such loss or damage shall be disregarded in computing any amounts payable to the Contractor under this Contract.

continued C·4

·3 The Contractor, for himself and for all his sub-contractors who pursuant to clause 6·9 are recognised as an insured under the Joint Names Policy referred to in paragraph C·2, shall authorise the insurers to pay all monies from such insurance in respect of the loss or damage to the Employer.

·4 If it is just and equitable, the Contractor's employment under this Contract may within 28 days of the occurrence of such loss or damage be terminated at the option of either Party by notice given to the other by actual delivery or by special or recorded delivery. If such notice is given:

   ·1 either Party may within 7 days of receiving such a notice (but not thereafter) invoke the dispute resolution procedures that apply under this Contract in order that it may be decided whether the termination is just and equitable; and

   ·2 upon the giving of such notice of termination or, where those dispute resolution procedures have been invoked, upon any final upholding of the notice of termination, the provisions of clauses 8·12·2 to 8·12·5 (except clause 8·12·3·5) shall apply.

·5 If no notice of termination is served under paragraph C·4·4, or if the notice of termination is disputed and is not upheld, then:

   ·1 after any inspection required by the insurers under the Joint Names Policy referred to in paragraph C·2 has been completed, the Contractor with due diligence shall restore the damaged work, replace or repair any lost or damaged Site Materials, remove and dispose of any debris and proceed with the carrying out and completion of the Works; and

   ·2 the restoration, replacement or repair of such loss or damage and (when required) the removal and disposal of debris shall be treated as a Variation.

The purpose of the Code is to assist in the fair and reasonable operation of the requirements of clause 3·18·4.

The Architect/Contract Administrator and the Contractor should endeavour to agree the amount and method of opening up or testing, but in any case, in issuing his instructions pursuant to that clause, the Architect/Contract Administrator is required to consider the following criteria:

1   the need in the event of non-compliance to demonstrate at no cost to the Employer either that it is unique and not likely to occur in similar elements of the Works or alternatively, the extent of any similar non-compliance in the Works already constructed or still to be constructed;

2   the need to discover whether any non-compliance in a primary structural element is a failure of workmanship and/or materials such that rigorous testing of similar elements must take place; or, where the non-compliance is in a less significant element, whether it is such as is to be statistically expected and can simply be repaired; or whether the non-compliance indicates an inherent weakness such as can only be found by selective testing, the extent of which must depend upon the importance of any detail concerned;

3   the significance of the non-compliance, having regard to the nature of the work in which it has occurred;

4   the consequence of any similar non-compliance on the safety of the building, its effect on users, adjoining property, the public, and compliance with any Statutory Requirements;

5   the level and standard of supervision and control of the Works by the Contractor;

6   the relevant records of the Contractor and, where relevant, those of any sub-contractor, whether resulting from the supervision and control referred to in paragraph 5 or otherwise;

7   any Codes of Practice or similar advice issued by a responsible body which are applicable to the non-compliant work, materials or goods;

8   any failure by the Contractor to carry out, or to secure the carrying out of, any tests specified in the Contract Documents or in an instruction of the Architect/Contract Administrator;

9   the reason for the non-compliance, when this has been established;

10   any technical advice that the Contractor has obtained in respect of the non-compliant work, materials or goods;

11   current recognised testing procedures;

12   the practicability of progressive testing in establishing whether any similar non-compliance is reasonably likely;

13   if alternative testing methods are available, the time required for and the consequential costs of such alternative testing methods;

14   any proposals of the Contractor; and

15   any other relevant matters.

Preliminary – Definitions

The terms 'the Consultants' and 'the Sub-Contractors' shall for the purposes of this Schedule mean the persons respectively identified as such in or by the P&T Rights Particulars or the Funder Rights Particulars, as the case may be.

Part 1: Third Party Rights for Purchasers and Tenants

('P&T Rights')

1 ·1 The Contractor warrants as at and with effect from practical completion of the Works (or, where there are Sections, practical completion of the relevant Section) that he has carried out the Works or, as the case may be, that Section, in accordance with this Contract. In the event of any breach of this warranty and subject to paragraphs 1·2, 1·3 and 1·4:

   ·1 the Contractor shall be liable for the reasonable costs of repair, renewal and/or reinstatement of any part or parts of the Works to the extent that the Purchaser or Tenant incurs such costs and/or the Purchaser or Tenant is or becomes liable either directly or by way of financial contribution for such costs; and

   ·2 (if paragraph 1·1·2 is stated in the P&T Rights Particulars to apply) the Contractor shall in addition to the costs referred to in paragraph 1·1·1 be liable for any other losses incurred by the Purchaser or Tenant up to the maximum liability stated in or by the P&T Rights Particulars.

·2 If in or by the P&T Rights Particulars paragraph 1·1·2 is stated or deemed not to apply, the Contractor shall not be liable for any losses incurred by the Purchaser or Tenant other than the costs referred to in paragraph 1·1·1.

·3 The Contractor's liability to a Purchaser or Tenant in respect of its P&T Rights shall be limited to the proportion of the Purchaser's or Tenant's losses which it would be just and equitable to require the Contractor to pay having regard to the extent of the Contractor's responsibility for the same, on the following assumptions, namely that:

   ·1 the Consultant(s) has or have provided contractual undertakings to or conferred third party rights on the Purchaser or Tenant as regards the performance of his or their services in connection with the Works in accordance with the terms of his or their respective consultancy agreements and that there are no limitations on liability as between the Consultant and the Employer in the consultancy agreement(s);

   ·2 the Sub-Contractor(s) has or have provided contractual undertakings to or conferred third party rights on the Purchaser or Tenant in respect of design of the Sub-Contract Works that he or they has or have carried out and for which there is no liability of the Contractor to the Employer under this Contract; and

   ·3 that the Consultant(s) and the Sub-Contractor(s) have paid to the Purchaser or Tenant such proportion of the Purchaser's or Tenant's losses as it would be just and equitable for them to pay having regard to the extent of their responsibility for the Purchaser's or Tenant's losses.

·4 The Contractor shall be entitled in any action or proceedings by the Purchaser or Tenant to rely on any term in this Contract and to raise the equivalent rights in defence of liability as he would have against the Employer under this Contract.

continued 1

·5 The obligations of the Contractor under or pursuant to this paragraph 1 shall not be released or diminished by the appointment of any person by the Purchaser or Tenant to carry out any independent enquiry into any relevant matter.

2 The Contractor further warrants that unless required by this Contract or unless otherwise authorised in writing by the Employer or by the Architect/Contract Administrator named in or appointed pursuant to this Contract (or, where such authorisation is given orally, confirmed in writing by the Contractor to the Employer and/or the Architect/Contract Administrator), he has not used and will not use materials in the Works other than in accordance with the guidelines contained in the edition of 'Good Practice in Selection of Construction Materials' (Ove Arup & Partners) current at the date of this Contract. In the event of any breach of this warranty the provisions of paragraph 1 shall apply.

3 The Purchaser or Tenant has no authority to issue any direction or instruction to the Contractor in relation to this Contract.

4 Where the Works include a Contractor's Designed Portion, the Purchaser or Tenant, insofar as it is the purchaser or tenant of any part(s) of the site falling within the Contractor's Designed Portion, and subject to the Contractor having been paid all monies due and payable under this Contract, shall in respect of such parts have rights and licences in relation to the Contractor's Design Documents in the same terms as those conferred on the Employer by clause 2·41, but subject to similar conditions, limitations and exclusions as apply thereunder to the Employer.

5 Where the Works include a Contractor's Designed Portion and this Contract so provides, the Contractor warrants that he has and shall maintain Professional Indemnity insurance in and on the terms and for the period referred to in clause 6·11 and its related Contract Particulars[66]. The Contractor shall immediately give notice to the Purchaser or Tenant if such insurance ceases to be available at commercially reasonable rates in order that the Contractor and the Purchaser or Tenant can discuss the means of best protecting their respective positions in the absence of such insurance. As and when it is reasonably requested to do so by the Purchaser or Tenant the Contractor shall produce for inspection documentary evidence that his Professional Indemnity insurance is being maintained.

6 P&T Rights may be assigned without the consent of the Contractor by a Purchaser or Tenant, by way of absolute legal assignment, to another person (P1) taking an assignment of the Purchaser's or Tenant's interest in the Works and by P1, by way of absolute legal assignment, to another person (P2) taking an assignment of P1's interest in the Works. In such cases the assignment shall only be effective upon written notice thereof being given to the Contractor. No further or other assignment of a Purchaser's or Tenant's rights under this Schedule will be permitted and in particular P2 shall not be entitled to assign these rights.

7 Any notice to be given by the Contractor shall be deemed to be duly given if it is delivered by hand or sent by special delivery or recorded delivery to the Purchaser or Tenant at its registered office; and any notice given by the Purchaser or Tenant shall be deemed to be duly given if it is delivered by hand or sent by special delivery or recorded delivery to the Contractor at its registered office; and in the case of any such notice, the same shall, if sent by special delivery or recorded delivery, be deemed (subject to proof to the contrary) to have been received 48 hours after being posted.

8 No action or proceedings for any breach of P&T Rights shall be commenced against the Contractor after the expiry of the relevant period from the date of practical completion of the Works. Where there are Sections no action or proceedings shall be commenced against the Contractor in respect of any Section after the expiry of the relevant period from the date of practical completion of such Section. For the purposes of this paragraph, the relevant period shall be:

·1 where this Contract is executed under hand, 6 years; and

·2 where this Contract is executed as a deed, 12 years.

9 For the avoidance of doubt, the Contractor shall have no liability to the Purchaser or Tenant under this Schedule for delay in completion of the Works.

---

[66] For Contractors who do not carry Professional Indemnity insurance, see the Guide.

10 This Schedule shall be governed by and construed in accordance with the law of England and the English courts shall have jurisdiction over any dispute or difference between the Contractor and any Purchaser or Tenant which arises out of or in connection with the P&T Rights of that Purchaser or Tenant.

Part 2: Third Party Rights for a Funder

('Funder Rights')

1 The Contractor warrants that he has complied and will continue to comply with this Contract. In the event of any breach of this warranty:

·1 the Contractor's liability to the Funder for costs under this Schedule shall be limited to the proportion of the Funder's losses which it would be just and equitable to require the Contractor to pay having regard to the extent of the Contractor's responsibility for the same, on the following assumptions, namely that:

·1 the Consultant(s) has or have provided contractual undertakings to or conferred third party rights on the Funder that he or they has or have and will perform his or their services in connection with the Works in accordance with the terms of his or their respective consultancy agreements and that there are no limitations on liability as between the Consultant and the Employer in the consultancy agreement(s);

·2 the Sub-Contractor(s) has or have provided contractual undertakings to or conferred third party rights on the Funder in respect of design of the Sub-Contract Works that he or they has or have carried out and for which there is no liability of the Contractor to the Employer under this Contract;

·3 the Consultant(s) and the Sub-Contractor(s) have paid to the Funder such proportion of the Funder's losses as it would be just and equitable for them to pay having regard to the extent of their responsibility for the Funder's losses;

·2 the Contractor shall be entitled in any action or proceedings by the Funder to rely on any term in this Contract and to raise the equivalent rights in defence of liability as he would have against the Employer under this Contract;

·3 the obligations of the Contractor under or pursuant to this paragraph 1 shall not be released or diminished by the appointment of any person by the Funder to carry out any independent enquiry into any relevant matter.

2 The Contractor further warrants that unless required by this Contract or unless otherwise authorised in writing by the Employer or by the Architect/Contract Administrator named in or appointed pursuant to this Contract (or, where such authorisation is given orally, confirmed in writing by the Contractor to the Employer and/or the Architect/Contract Administrator), he has not used and will not use materials in the Works other than in accordance with the guidelines contained in the edition of 'Good Practice in Selection of Construction Materials' (Ove Arup & Partners) current at the date of this Contract. In the event of any breach of this warranty the provisions of paragraph 1 shall apply.

3 The Funder has no authority to issue any direction or instruction to the Contractor in relation to this Contract unless and until the Funder has given notice under paragraph 5 or 6·4.

4 The Funder has no liability to the Contractor in respect of amounts due under this Contract unless and until the Funder has given notice under paragraph 5 or 6·4.

5 The Contractor agrees that, in the event of the termination of the Finance Agreement by the Funder, the Contractor shall, if so required by notice in writing given by the Funder and subject to paragraph 7, accept the instructions of the Funder or its appointee to the exclusion of the Employer in respect of the Works upon the terms and conditions of this Contract. The Employer acknowledges that the Contractor shall be entitled to rely on a notice given to the Contractor by the Funder under this paragraph 5 as conclusive evidence for the purposes of this Contract of the termination of the Finance Agreement by the Funder; and further acknowledges that such acceptance of the instructions of the Funder to the exclusion of the Employer shall not constitute any breach of the Contractor's obligations to the Employer under this Contract.

6 **·1** The Contractor shall not exercise any right of termination of his employment under this Contract without having first:

    ·1 copied to the Funder any written notices required by this Contract to be sent to the Architect/ Contract Administrator or to the Employer prior to the Contractor being entitled to give notice under this Contract that his employment under this Contract is terminated; and

    ·2 given to the Funder written notice that he has the right under this Contract forthwith to notify the Employer that his employment under this Contract is terminated.

  ·2 The Contractor shall not treat this Contract as having been repudiated by the Employer without having first given to the Funder written notice that he intends so to inform the Employer.

  ·3 The Contractor shall not:

    ·1 issue any notification to the Employer to which paragraph 6·1·2 refers; or

    ·2 inform the Employer that he is treating this Contract as having been repudiated by the Employer as referred to in paragraph 6·2

    before the lapse of 7 days (or such other period as may be stated in the Funder Rights Particulars) from receipt by the Funder of the written notice by the Contractor which the Contractor is required to give under paragraph 6·1·2 or 6·2.

  ·4 The Funder may, not later than the expiry of the period referred to in paragraph 6·3, require the Contractor by notice in writing and subject to paragraph 7 to accept the instructions of the Funder or its appointee to the exclusion of the Employer in respect of the Works upon the terms and conditions of this Contract. The Employer acknowledges that the Contractor shall be entitled to rely on a notice given to the Contractor by the Funder under this paragraph 6·4 and that acceptance by the Contractor of the instructions of the Funder to the exclusion of the Employer shall not constitute any breach of the Contractor's obligations to the Employer under this Contract. Provided that, subject to paragraph 7, nothing in this paragraph 6·4 shall relieve the Contractor of any liability he may have to the Employer for any breach by the Contractor of this Contract.

7 It shall be a condition of any notice given by the Funder under paragraph 5 or 6·4 that the Funder or its appointee accepts liability for payment of the sums due and payable to the Contractor under this Contract and for performance of the Employer's obligations including payment of any sums outstanding at the date of such notice. Upon the issue of any notice by the Funder under paragraph 5 or 6·4, this Contract shall continue in full force and effect as if no right of termination of the Contractor's employment under this Contract, nor any right of the Contractor to treat this Contract as having been repudiated by the Employer, had arisen and the Contractor shall be liable to the Funder and its appointee under this Contract in lieu of his liability to the Employer. If any notice given by the Funder under paragraph 5 or 6·4 requires the Contractor to accept the instructions of the Funder's appointee, the Funder shall be liable to the Contractor as guarantor for the payment of all sums from time to time due to the Contractor from the Funder's appointee.

8 Where the Works include a Contractor's Designed Portion and subject to the Contractor having been paid all monies due and payable to under this Contract, the Funder shall in respect of such parts have rights and licences in relation to the Contractor's Design Documents in the same terms as those conferred on the Employer by clause 2·41, but subject to similar conditions, limitations and exclusions as apply thereunder to the Employer.

9 Where the Works include a Contractor's Designed Portion the Contractor has and shall maintain Professional Indemnity insurance in and on the terms and for the period referred to in clause 6·11 and its related Contract Particulars[66]. The Contractor shall immediately give notice to the Funder if such insurance ceases to be available at commercially reasonable rates in order that the Contractor and the Funder can discuss the means of best protecting their respective positions in the absence of such insurance. As and when it is reasonably requested to do so by the Funder or its appointee under paragraph 5 or 6·4 the Contractor shall produce for inspection documentary evidence that his Professional Indemnity insurance is being maintained.

10   The rights contained in this Schedule may be assigned without the consent of the Contractor by the Funder, by way of absolute legal assignment, to another person (P1) providing finance or re-finance in connection with the carrying out of the Works and by P1, by way of absolute legal assignment, to another person (P2) providing finance or re-finance in connection with the carrying out of the Works. In such cases the assignment shall only be effective upon written notice thereof being given to the Contractor. No further or other assignment of Funder Rights will be permitted and in particular P2 shall not be entitled to assign these rights.

11   Any notice to be given by the Contractor shall be deemed to be duly given if it is delivered by hand or sent by special delivery or recorded delivery to the Funder at its registered office; and any notice given by the Funder shall be deemed to be duly given if it is delivered by hand or sent by special delivery or recorded delivery to the Contractor at its registered office; and in the case of any such notices, the same shall, if sent by special delivery or recorded delivery, be deemed (subject to proof to the contrary) to have been received 48 hours after being posted.

12   No action or proceedings for any breach of the rights contained in this Schedule shall be commenced against the Contractor after the expiry of the relevant period from the date of practical completion of the Works. Where there are Sections no action or proceedings shall be commenced against the Contractor in respect of any Section after the expiry of the relevant period from the date of practical completion of such Section. For the purposes of this paragraph, the relevant period shall be:

    ·1   where this Contract is executed under hand, 6 years; and

    ·2   where this Contract is executed as a deed, 12 years.

13   Notwithstanding the rights contained in this Schedule, the Contractor shall have no liability to the Funder for delay under this Contract unless and until the Funder serves notice pursuant to paragraph 5 or 6·4. For the avoidance of doubt the Contractor shall not be required to pay liquidated damages in respect of the period of delay where the same has been paid to or deducted by the Employer.

14   ·1   This Schedule shall be governed by and construed in accordance with the law of England and subject to paragraph 14·2 the English courts shall have jurisdiction over any dispute or difference between the Contractor and the Funder which arises out of or in connection with this Schedule.

    ·2   Following the giving of any notice by the Funder pursuant to paragraph 5 or 6·4, any dispute or difference which shall arise between the Contractor and the Funder (including any appointee or permitted assignee) shall be subject to the provisions of Article 7 and (where they apply) Article 8 and clauses 9·3 to 9·8.

(Agreed between the JCT and the British Bankers' Association)

Part 1: Advance Payment Bond[67]

1 THE parties to this Bond are:

_____

whose registered office is at _____

_____ ('the Surety'), and

_____

of _____

_____ ('the Employer').

2 The Employer and _____ ('the Contractor')

have agreed to enter into a contract ('the Contract') for building works ('the Works') at _____

_____

3 The Employer has agreed to pay the Contractor the sum of [_____] as an advance payment of sums due to the Contractor under the Contract ('the Advance Payment') for reimbursement by the Surety on the following terms:

·1 when the Surety receives a demand from the Employer in accordance with clause 3·2 below the Surety shall repay the Employer the sum demanded up to the amount of the Advance Payment;

·2 the Employer shall in making any demand provide to the Surety a completed notice of demand in the form of the **Schedule** attached hereto which shall be accepted as conclusive evidence for all purposes under this Bond. The signatures on any such demand must be authenticated by the Employer's bankers;

·3 the Surety shall within 5 Business Days after receiving the demand pay to the Employer the sum so demanded. 'Business Day' means the day (other than a Saturday or a Sunday) on which commercial banks are open for business in London.

4 Payments due under this Bond shall be made notwithstanding any dispute between the Employer and the Contractor and whether or not the Employer and the Contractor are or might be under any liability one to the other. Payment by the Surety under this Bond shall be deemed a valid payment for all purposes of this Bond and shall discharge the Surety from liability to the extent of such payment.

5 The Surety consents and agrees that the following actions by the Employer may be made and done without notice to or consent of the Surety and without in any way affecting changing or releasing the Surety from its obligations under this Bond and the liability of the Surety hereunder shall not in any way be affected hereby. The actions are:

[67] Not applicable where the Employer is a Local Authority.

continued 5

·1 waiver by the Employer of any of the terms, provisions, conditions, obligations and agreements of the Contractor or any failure to make demand upon or take action against the Contractor;

·2 any modification or changes to the Contract; and/or

·3 the granting of any extensions of time to the Contractor without affecting the terms of clause 7·3 below.

6 The Surety's maximum aggregate liability under this Bond which shall commence on payment of the Advance Payment by the Employer to the Contractor shall be the amount of [_____] which sum shall be reduced by the amount of any reimbursement made by the Contractor to the Employer as advised by the Employer in writing to the Surety.

7 The obligations of the Surety under this Bond shall cease upon whichever is the earliest of:

·1 the date on which the Advance Payment is reduced to nil as certified in writing to the Surety by the Employer;

·2 the date on which the Advance Payment or any balance thereof is repaid to the Employer by the Contractor (as certified in writing to the Surety by the Employer) or by the Surety; and

·3 [*longstop date to be given*],

and any claims hereunder must be received by the Surety in writing on or before such earliest date.

8 This Bond is not transferable or assignable without the prior written consent of the Surety. Such written consent will not be unreasonably withheld.

9 Notwithstanding any other provisions of this Bond nothing in this Bond confers or is intended to confer any right to enforce any of its terms on any person who is not a party to it.

10 This Bond shall be governed and construed in accordance with the laws of England and Wales.

IN WITNESS whereof this Deed of Guarantee has been duly executed and delivered on the date below:

Signed as a Deed by: _____

as the Attorney and on behalf of the Surety: _____

In the presence of:

_____
witness' signature

_____
witness' name

_____
witness' address

_____

Date: _____

Schedule to Advance Payment Bond

(clause 3·2 of the Bond)

Notice of Demand

Date of Notice: _____

Date of Bond: _____

Employer: _____

Surety: _____

The Bond has come into effect.

We hereby demand payment of the sum of

£ _____(amount in words)
which does not exceed the amount of reimbursement for which the Contractor is in default at the date of this notice.

Address for payment: _____

_____

This Notice is signed by the following persons who are authorised by the Employer to act for and on his behalf:

Signed by _____

    Name: _____

    Official Position: _____

Signed by_____

    Name: _____

    Official Position: _____

The above signatures to be authenticated by the Employer's bankers

Part 2: Bond in respect of payment for off-site materials and/or goods

1   THE parties to this Bond are:

_____

whose registered office is at _____

_____ ('the Surety'), and

_____

of _____

_____ ('the Employer').

2   The Employer and _____ ('the Contractor')

have agreed to enter into a contract ('the Contract') for building works ('the Works') at _____

_____

3   Subject to the relevant provisions of the Contract as summarised below but with which the Surety shall not at all be concerned:

    ·1   the Employer has agreed to include the amount stated as due in Interim Certificates (as defined in the Contract) for payment by the Employer the value of those materials or goods or items pre-fabricated for inclusion in the Works listed by the Employer in a list which has been included as part of the Contract ('the Listed Items'), before their delivery to or adjacent to the Works; and

    ·2   the Contractor has agreed to insure the Listed Items against loss or damage for their full value under a policy of insurance protecting the interests of the Employer and the Contractor during the period commencing with the transfer of the property in the items to the Contractor until they are delivered to or adjacent to the Works; and

    ·3   this Bond shall exclusively relate to the amount paid to the Contractor in respect of the Listed Items which have not been delivered to or adjacent to the Works.

4   The Employer shall in making any demand provide to the Surety a Notice of Demand in the form of the **Schedule** attached hereto which shall be accepted as conclusive evidence for all purposes under this Bond. The signatures on any such demand must be authenticated by the Employer's bankers.

5   The Surety shall within 5 Business Days after receiving the demand pay to the Employer the sum so demanded. 'Business Day' means the day (other than a Saturday or a Sunday) on which commercial banks are open for business in London.

6   Payments due under this Bond shall be made notwithstanding any dispute between the Employer and the Contractor and whether or not the Employer and the Contractor are or might be under any liability one to the other. Payment by the Surety under this Bond shall be deemed a valid payment for all purposes of this Bond and shall discharge the Surety from liability to the extent of such payment.

7   The Surety consents and agrees that the following actions by the Employer may be made and done without notice to or consent of the Surety and without in any way affecting changing or releasing the Surety from its obligations under this Bond and the liability of the Surety hereunder shall not in any way be affected hereby. The actions are:

    ·1   waiver by the Employer of any of the terms, provisions, conditions, obligations and agreements of the Contractor or any failure to make demand upon or take action against the Contractor;

continued 7

·2   any modification or changes to the Contract; and/or

·3   the granting of an extension of time to the Contractor without affecting the terms of clause 9·2 below.

8   The Surety's maximum aggregate liability under this Bond shall be *[_____].

9   The obligations of the Surety under this Bond shall cease upon whichever is the earlier of:

·1   the date on which all the Listed Items have been delivered to or adjacent to the Works as certified in writing to the Surety by the Employer; or

·2   [*longstop date to be given*],

and any claims hereunder must be received by the Surety in writing on or before such earlier date.

10   The Bond is not transferable or assignable without the prior written consent of the Surety. Such written consent will not be unreasonably withheld.

11   Notwithstanding any other provisions of this Bond nothing in this Bond confers or is intended to confer any right to enforce any of its terms on any person who is not a party to it.

12   This Bond shall be governed and construed in accordance with the laws of England and Wales.

*The value stated in the Contract which the Employer considers will be sufficient to cover him for maximum payments to the Contractor for the Listed Items that will have been made and not delivered to the site at any one time.

IN WITNESS whereof this Deed of Guarantee has been duly executed and delivered on the date below:

Signed as a Deed by: _____

as the Attorney and on behalf of the Surety: _____

In the presence of:

_____
witness' signature

_____
witness' name

_____
witness' address

Date: _____

Schedule to Bond

(clause 4 of the Bond)

Notice of Demand

Date of Notice: _____

Date of Bond: _____

Employer: _____

Surety: _____

We hereby demand payment of the sum of £ _____
being the amount stated as due in respect of Listed Items included in the amount stated as due in an Interim Certificate(s) for payment which has been duly made to the Contractor by the Employer but such Listed Items have not been delivered to or adjacent to the Works.

Address for payment: _____

_____

This Notice is signed by the following persons who are authorised by the Employer to act for and on his behalf:

Signed by _____

      Name: _____

      Official Position: _____

Signed by _____

      Name: _____

      Official Position: _____

The above signatures to be authenticated by the Employer's bankers

Part 3: Retention Bond[68]

BOND dated the _____ day of _____ 20 _____

issued by _____

of _____

_____ ('the Surety')

in favour of _____

of _____

_____ ('the Employer')

1   By a contract ('the Contract') between the Employer and

_____

of _____

_____ ('the Contractor')
the Employer has agreed that he will not exercise his right under the Contract to deduct Retention from amounts included in Interim Certificates provided the Contractor has taken out this Bond in favour of the Employer.

2   The Surety is hereby bound to the Employer in the maximum aggregate sum of _____

_____ (figures and words)
until the Surety is notified by the Employer in writing of the date of issue of the next Interim Certificate after practical completion when the maximum aggregate sum shall be reduced by 50 per cent.

3   The Employer shall, on a demand which complies with the requirements in clause 4 below, be entitled to receive from the Surety the sum therein demanded.

4   Any demand by the Employer under clause 3 above shall:

    ·1   be in writing addressed to the Surety at its office at

    _____

    refer to this Bond, and with the signature(s) therein authenticated by the Employer's bankers; and

    ·2   state the amount of the Retention that would have been held by the Employer at the date of the demand had Retention been deductible; and

    ·3   state the amount demanded, which shall not exceed the amount stated pursuant to clause 4·2 above, and identify for which one or more of the following such amount is demanded:

---

[68]   Not applicable where the Employer is a Local Authority.

continued 4·3

·1 the costs actually incurred by the Employer by reason of the failure of the Contractor to comply with the instructions of the Architect/Contract Administrator under the Contract; and be accompanied by a statement by the Architect/Contract Administrator which confirms that this failure by the Contractor has occurred;

·2 the insurance premiums paid by the Employer pursuant to the Contract because the Contractor has not taken out and/or not maintained any insurance of the building works which he was required under the Contract to take out and/or maintain;

·3 liquidated and ascertained damages which under the Contract the Contractor is due to pay or allow to the Employer; and be accompanied by a copy of the certificate of the Architect/Contract Administrator which under the Contract he is required to issue and which certifies that the Contractor has failed to complete the works by the contractual Completion Date;

·4 any expenses or any direct loss and/or damage caused to the Employer as a result of the termination of the Contractor's employment by the Employer;

·5 any costs, other than the amounts referred to in clauses 4·3·1 to 4·3·4 above, which the Employer has actually incurred and which, under the Contract, he is entitled to deduct from monies otherwise due or to become due to the Contractor; and identify his entitlement;

and

·4 incorporate a certification that the Contractor has been given 14 days' written notice of his liability for the amount demanded hereunder by the Employer and that the Contractor has not discharged that liability; and that a copy of this notice has at the same time been sent to the Surety at its office at

_____

Such demand as above shall, for the purposes of this Bond but not further or otherwise, be conclusive evidence (and admissible as such) that the amount demanded is properly due and payable to the Employer by the Contractor.

5 If the Contract is to be assigned or otherwise transferred with the benefit of this Bond, the Employer shall be entitled to assign or transfer this Bond only with the prior written consent of the Surety, such consent not to be unreasonably delayed or withheld.

6 The Surety, in the absence of a prior written demand made, shall be released from its liability under this Bond upon the earliest occurrence of either:

·1 the date of issue under the Contract of the Certificate of Making Good as confirmed by the Employer to the Surety; or

·2 satisfaction of a demand(s) up to the maximum aggregate under the Bond; or

·3 _____ (insert calendar date).

7 Any demand made hereunder must be received by the Surety accompanied by the documents as required by clause 4 above on or before the earliest occurrence as stated above, when this Bond will terminate and become of no further effect whatsoever.

8 Notwithstanding any other provisions of this Bond nothing in this Bond confers or is intended to confer any right to enforce any of its terms on any person who is not a party to it.

9 This Bond shall be governed and construed in accordance with the laws of England and Wales.

IN WITNESS whereof this Deed of Guarantee has been duly executed and delivered on the date below:

Signed as a Deed by: _____

as the Attorney and on behalf of the Surety: _____

In the presence of:

_____
witness' signature

_____
witness' name

_____
witness' address

_____

Date: _____

Notes[69]

1   The terms of this Retention Bond have been agreed with the British Bankers' Association and discussed with the Surety Panel of the Association of British Insurers. The JCT understands that a Bond which embodies the terms of this Part 3 of Schedule 6 is, at the proposed Surety's discretion, available to a Contractor where the Employer has incorporated into a building contract in the JCT Standard Building Contract 2005 Edition, optional clause 4·19.

2   In clause 2 the figure to be inserted here is the amount stated in the Contract Particulars pursuant to clause 4·19. It is understood that a Surety will, at additional cost to the Contractor, and which may be subject to other terms and conditions of the Surety, provide for a greater sum than that stated in clause 2 of the Bond if, due to variations, and had Retention been applicable, that amount would have increased. The reduction by 50% of the maximum aggregate sum at the date of issue of the next Interim Certificate after practical completion matches a similar reduction had Retention been applicable.

3   The inclusion in the last paragraph of clause 4 of the words "but not further or otherwise" is to make clear that the Contractor would not be prevented by the terms of clause 4 from alleging, under the Contract, that the Contractor was not in breach on any of the matters stated in clauses 4·3·1 to 4·3·5 of the Bond.

Any demand by the Employer under clause 4 of the Bond must not exceed the costs actually incurred by the Employer and is not to be in excess of the amount stated pursuant to clause 4·2.

4   The Surety will require an actual expiry date to be stated in clause 6·3 of the Bond or (if earlier) a date that is capable of being ascertained on the face of the Bond. Where this is not possible, alternative terms should be discussed with the Surety.

[69]   These notes will not appear on the Bond issued by the Surety.

Fluctuations Option A

Contribution, levy and tax fluctuations

Deemed calculation of Contract Sum – labour

A·1   The Contract Sum shall be deemed to have been calculated in the manner set out below and shall be subject to adjustment in the events specified hereunder.

·1   The Contract Sum is based upon the types and rates of contribution, levy and tax payable by a person in his capacity as an employer and which at the Base Date are payable by the Contractor. A type and a rate so payable are in paragraph A·1·2 referred to as a 'tender type' and a 'tender rate'.

·2   If any of the tender rates other than a rate of levy payable by virtue of the Industrial Training Act 1982 is increased or decreased, or if a tender type ceases to be payable, or if a new type of contribution, levy or tax which is payable by a person in his capacity as an employer becomes payable after the Base Date, then in any such case the net amount of the difference between what the Contractor actually pays or will pay in respect of:

·1   workpeople engaged upon or in connection with the Works either on or adjacent to the site; and

·2   workpeople directly employed by the Contractor who are engaged upon the production of materials or goods for use in or in connection with the Works and who operate neither on nor adjacent to the site and to the extent that they are so engaged

or because of his employment of such workpeople and what he would have paid had the alteration, cessation or new type of contribution, levy or tax not become effective shall, as the case may be, be paid to or allowed by the Contractor.

·3   There shall be added to the net amount paid to or allowed by the Contractor under paragraph A·1·2, in respect of each person employed by the Contractor who is engaged upon or in connection with the Works either on or adjacent to the site and who is not within the definition of workpeople in paragraph A·11·3, the same amount as is payable or allowable in respect of a craftsman under paragraph A·1·2 or such proportion of that amount as reflects the time (measured in whole working days) that each such person is so employed.

·4   For the purposes of paragraph A·1·3:

·1   no period of less than 2 whole working days in any week shall be taken into account and periods of less than a whole working day shall not be aggregated to amount to a whole working day;

·2   "the same amount as is payable or allowable in respect of a craftsman" shall refer to the amount in respect of a craftsman employed by the Contractor (or by any sub-contractor under a sub-contract to which paragraph A·3 refers) under the rules or decisions or agreements of the Construction Industry Joint Council or other wage-fixing body and, where those rules or decisions or agreements provide for more than one rate of wage, emolument or other expense for a craftsman, shall refer to the amount in respect of a craftsman employed as aforesaid to whom the highest rate is applicable; and

·3   "employed by the Contractor" shall mean an employment to which the Income Tax (Pay As You Earn) Regulations 2003 apply.

·5   The Contract Sum is based upon the types and rates of refund of the contributions, levies and taxes payable by a person in his capacity as an employer and upon the types and rates of premium receivable by a person in his capacity as an employer being in each case types and rates which at

continued A·1·5
the Base Date are receivable by the Contractor. Such a type and such a rate are in paragraph A·1·6 referred to as a 'tender type' and a 'tender rate'.

·6   If any of the tender rates is increased or decreased or if a tender type ceases to be payable or if a new type of refund of any contribution, levy or tax payable by a person in his capacity as an employer becomes receivable or if a new type of premium receivable by a person in his capacity as an employer becomes receivable after the Base Date, then in any such case the net amount of the difference between what the Contractor actually receives or will receive in respect of workpeople as referred to in paragraphs A·1·2·1 and A·1·2·2 or because of his employment of such workpeople and what he would have received had the alteration, cessation or new type of refund or premium not become effective shall, as the case may be, be paid to or allowed by the Contractor.

·7   The references in paragraphs A·1·5 and A·1·6 to premiums shall be construed as meaning all payments howsoever they are described which are made under or by virtue of an Act of Parliament to a person in his capacity as an employer and which affect the cost to an employer of having persons in his employment.

·8   Where employer's contributions are payable by the Contractor in respect of workpeople as referred to in paragraphs A·1·2·1 and A·1·2·2 whose employment is contracted-out employment within the meaning of the Pension Schemes Act 1993, the Contractor shall for the purpose of recovery or allowance under this paragraph A·1 be deemed to pay employer's contributions as if that employment were not contracted-out employment.

·9   The references in paragraph A·1 to contributions, levies and taxes shall be construed as meaning all impositions payable by a person in his capacity as an employer howsoever they are described and whoever the recipient which are imposed under or by virtue of an Act of Parliament and which affect the cost to an employer of having persons in his employment.

Deemed calculation of Contract Sum – materials

A·2   The Contract Sum shall be deemed to have been calculated in the manner set out below and shall be subject to adjustment in the events specified hereunder.

·1   The Contract Sum is based upon the types and rates of duty, if any, and tax, if any (other than any VAT which is treated, or is capable of being treated, as input tax by the Contractor), by whomsoever payable which at the Base Date are payable on the import, purchase, sale, appropriation, processing, use or disposal of the materials, goods, electricity, fuels, materials taken from the site as waste or any other solid, liquid or gas necessary for the execution of the Works by virtue of any Act of Parliament. A type and a rate so payable are in paragraph A·2·2 referred to as a 'tender type' and a 'tender rate'.

·2   If, in relation to any materials or goods or any electricity or fuels or materials taken from the site as waste or any other solid, liquid or gas necessary for the execution of the Works including temporary site installations for those Works, a tender rate is increased or decreased or a tender type ceases to be payable or a new type of duty or tax (other than any VAT which is treated, or is capable of being treated, as input tax by the Contractor) becomes payable on the import, purchase, sale, appropriation, processing, use or disposal of any of the above things after the Base Date, then in any such case the net amount of the difference between what the Contractor actually pays in respect of those materials, goods, electricity, fuels, materials taken from the site as waste or any other solid, liquid or gas and what he would have paid in respect of them had the alteration, cessation or imposition not occurred shall, as the case may be, be paid to or allowed by the Contractor. In this paragraph A·2·2 "a new type of duty or tax" includes an additional duty or tax and a duty or tax imposed in regard to any of the above in respect of which no duty or tax whatever was previously payable (other than any VAT which is treated, or is capable of being treated, as input tax by the Contractor).

Sub-let work – incorporation of provisions to like effect

A·3   ·1   If the Contractor sub-lets any portion of the Works to a sub-contractor he shall incorporate in the sub-contract provisions to the like effect as the provisions of Fluctuations Option A (excluding this paragraph A·3) including the percentage stated in the Contract Particulars pursuant to paragraph A·12 which are applicable for the purposes of this Contract.

·2   If the price payable under such a sub-contract as referred to in paragraph A·3·1 is increased above or decreased below the price in such sub-contract by reason of the operation of the said incorporated *continued A·3·2*   provisions, then the net amount of such increase or decrease shall, as the case may be, be paid to or allowed by the Contractor under this Contract.

Written notice by Contractor

A·4   ·1   The Contractor shall give a written notice to the Architect/Contract Administrator of the occurrence of any of the events referred to in such of the following provisions as are applicable for the purposes of this Contract:

·1   paragraph A·1·2;

·2   paragraph A·1·6;

·3   paragraph A·2·2;

·4   paragraph A·3·2.

·2   Any notice required to be given under paragraph A·4·1 shall be given within a reasonable time after the occurrence of the event to which the notice relates, and the giving of a written notice in that time shall be a condition precedent to any payment being made to the Contractor in respect of the event in question.

Agreement – Quantity Surveyor and Contractor

A·5   The Quantity Surveyor and the Contractor may agree what shall be deemed for all the purposes of this Contract to be the net amount payable to or allowable by the Contractor in respect of the occurrence of any event such as is referred to in any of the provisions listed in paragraph A·4·1.

Fluctuations added to or deducted from Contract Sum

A·6   Any amount which from time to time becomes payable to or allowable by the Contractor by virtue of paragraphs A·1 and A·2 or paragraph A·3 shall, as the case may be, be added to or deducted from:

·1   the Contract Sum; and

·2   any amounts payable to the Contractor and which are calculated in accordance with clause 8·12·3·1.

The addition or deduction to which this paragraph A·6 refers shall be subject to the provisions of paragraphs A·7 to A·9·1.

Evidence and computations by Contractor

A·7   As soon as is reasonably practicable the Contractor shall provide such evidence and computations as the Architect/Contract Administrator or the Quantity Surveyor may reasonably require to enable the amount payable to or allowable by the Contractor by virtue of paragraphs A·1 and A·2 or paragraph A·3 to be ascertained; and in the case of amounts payable to or allowable by the Contractor under paragraph A·1·3 (or paragraph A·3 for amounts payable to or allowable under the provisions in the sub-contract to the like effect as paragraphs A·1·3 and A·1·4) – employees other than workpeople – such evidence shall include a certificate signed by or on behalf of the Contractor each week certifying the validity of the evidence reasonably required to ascertain such amounts.

No alteration to Contractor's profit

A·8   No addition to or deduction from the Contract Sum made by virtue of paragraph A·6 shall alter in any way the amount of profit of the Contractor included in that Sum.

Position where Contractor in default over completion

A·9   ·1   Subject to the provisions of paragraph A·9·2 no amount shall be added or deducted in the computation of the amount stated as due in an Interim Certificate or in the Final Certificate in respect of amounts otherwise payable to or allowable by the Contractor by virtue of paragraphs A·1 and A·2 or paragraph A·3 if the event (as referred to in the provisions listed in paragraph A·4·1) in respect of which the payment or allowance would be made occurs after the Completion Date.

continued A·9

·2   Paragraph A·9·1 shall not be applied unless:

   ·1   the printed text of clauses 2·26 to 2·29 is unamended and forms part of the Conditions; and

   ·2   the Architect/Contract Administrator has, in respect of every written notification by the Contractor under clause 2·28, fixed or confirmed in writing such Completion Date as he considers to be in accordance with that clause.

Work etc. to which paragraphs A·1 to A·3 not applicable

A·10   Paragraphs A·1 to A·3 shall not apply in respect of:

·1   work for which the Contractor is allowed daywork rates under clause 5·7;

·2   changes in the rate of VAT charged on the supply of goods or services by the Contractor to the Employer under this Contract.

Definitions for use with Fluctuations Option A

A·11   In Fluctuations Option A:

·1   the Base Date means the date stated as such in the Contract Particulars;

·2   "materials" and "goods" include timber used in formwork but do not include other consumable stores, plant and machinery;

·3   "workpeople" means persons whose rates of wages and other emoluments (including holiday credits) are governed by the rules or decisions or agreements of the Construction Industry Joint Council or some other wage-fixing body for trades associated with the building industry;

·4   "wage-fixing body" means a body which lays down recognised terms and conditions of workers;

·5   "recognised terms and conditions" means terms and conditions of workers in comparable employment in the trade or industry, or section of trade or industry, in which the employer in question is engaged which have been settled by an agreement or award to which the parties are employers' associations and independent trade unions which represent (generally, or in the district in question, as the case may be) a substantial proportion of the employers and of the workers in the trade, industry or section being workers of the description to which the agreement or award relates.

Percentage addition to fluctuation payments or allowances

A·12   There shall be added to the amount paid to or allowed by the Contractor under:

·1   paragraph A·1·2,

·2   paragraph A·1·3,

·3   paragraph A·1·6,

·4   paragraph A·2·2

the percentage stated in the Contract Particulars.

Fluctuations Option B

Labour and materials cost and tax fluctuations

Deemed calculation of Contract Sum – labour rates etc.

B·1 The Contract Sum shall be deemed to have been calculated in the manner set out below and shall be subject to adjustment in the events specified hereunder.

·1 The Contract Sum (including the cost of employer's liability insurance and of third party insurance) is based upon the rates of wages and the other emoluments and expenses (including holiday credits) which will be payable by the Contractor to or in respect of:

·1 workpeople engaged upon or in connection with the Works either on or adjacent to the site; and

·2 workpeople directly employed by the Contractor who are engaged upon the production of materials or goods for use in or in connection with the Works and who operate neither on nor adjacent to the site and to the extent that they are so engaged

in accordance with:

·3 the rules or decisions of the Construction Industry Joint Council or other wage-fixing body which will be applicable to the Works and which have been promulgated at the Base Date;

·4 any incentive scheme and/or productivity agreement under the Working Rule Agreement of the Construction Industry Joint Council or provisions on incentive schemes and/or productivity agreements contained in the rules or decisions of some other wage-fixing body; and

·5 the terms of the Building and Civil Engineering Annual and Public Holidays Agreements (or the terms of agreements to similar effect in respect of workpeople whose rates of wages and other emoluments and expenses (including holiday credits) are in accordance with the rules or decisions of a wage-fixing body other than the Construction Industry Joint Council) which will be applicable to the Works and which have been promulgated at the Base Date;

and upon the rates or amounts of any contribution, levy or tax which will be payable by the Contractor in his capacity as an employer in respect of, or calculated by reference to, the rates of wages and other emoluments and expenses (including holiday credits) referred to herein.

·2 If any of the said rates of wages or other emoluments and expenses (including holiday credits) are increased or decreased by reason of any alteration in the said rules, decisions or agreements promulgated after the Base Date, then the net amount of the increase or decrease in wages or other emoluments and expenses (including holiday credits) together with the net amount of any consequential increase or decrease in the cost of employer's liability insurance, of third party insurance and of any contribution, levy or tax payable by a person in his capacity as an employer shall, as the case may be, be paid to or allowed by the Contractor.

·3 There shall be added to the net amount paid to or allowed by the Contractor under paragraph B·1·2, in respect of each person employed by the Contractor who is engaged upon or in connection with the Works either on or adjacent to the site and who is not within the definition of workpeople in paragraph B·12·3, the same amount as is payable or allowable in respect of a craftsman under paragraph B·1·2 or such proportion of that amount as reflects the time (measured in whole working days) that each such person is so employed.

·4 For the purposes of paragraphs B·1·3 and B·2·3:

·1 no period of less than 2 whole working days in any week shall be taken into account and periods of less than a whole working day shall not be aggregated to amount to a whole working day;

·2 "the same amount as is payable or allowable in respect of a craftsman" shall refer to the amount in respect of a craftsman employed by the Contractor (or by any sub-contractor under a sub-contract to which paragraph B·4 refers) under the rules or decisions or agreements of the Construction Industry Joint Council or other wage-fixing body and, where those rules or decisions or

continued B·1·4·2

agreements provide for more than one rate of wage, emolument or other expense for a craftsman, shall refer to the amount in respect of a craftsman employed as aforesaid to whom the highest rate is applicable; and

·3 "employed by the Contractor" shall mean an employment to which the Income Tax (Pay As You Earn) Regulations 2003 apply.

·5 The Contract Sum is based upon:

·1 the transport charges referred to in a basic transport charges list submitted by the Contractor and attached to the Contract Documents and incurred by the Contractor in respect of workpeople engaged in either of the capacities referred to in paragraphs B·1·1·1 and B·1·1·2; or

·2 the reimbursement of fares which will be reimbursable by the Contractor to workpeople engaged in either of the capacities referred to in paragraphs B·1·1·1 and B·1·1·2 in accordance with the rules or decisions of the Construction Industry Joint Council which will be applicable to the Works and which have been promulgated at the Base Date or, in the case of workpeople so engaged whose rates of wages and other emoluments and expenses are governed by the rules or decisions of some wage-fixing body other than the Construction Industry Joint Council, in accordance with the rules or decisions of such other body which will be applicable and which have been promulgated as aforesaid.

·6 If:

·1 the amount of transport charges referred to in the basic transport charges list is increased or decreased after the Base Date; or

·2 the reimbursement of fares is increased or decreased by reason of any alteration in the said rules or decisions promulgated after the Base Date or by any actual increase or decrease in fares which takes effect after the Base Date,

then the net amount of that increase or decrease shall, as the case may be, be paid to or allowed by the Contractor.

Deemed calculation of Contract Sum – labour levies and taxes

B·2 The Contract Sum shall be deemed to have been calculated in the manner set out below and shall be subject to adjustment in the events specified hereunder.

·1 The Contract Sum is based upon the types and rates of contribution, levy and tax payable by a person in his capacity as an employer and which at the Base Date are payable by the Contractor. A type and a rate so payable are in paragraph B·2·2 referred to as a 'tender type' and a 'tender rate'.

·2 If any of the tender rates other than a rate of levy payable by virtue of the Industrial Training Act 1982 is increased or decreased, or if a tender type ceases to be payable, or if a new type of contribution, levy or tax which is payable by a person in his capacity as an employer becomes payable after the Base Date, then in any such case the net amount of the difference between what the Contractor actually pays or will pay in respect of workpeople as referred to in paragraphs B·1·1·1 and B·1·1·2 or because of his employment of such workpeople and what he would have paid had the alteration, cessation or new type of contribution, levy or tax not become effective shall, as the case may be, be paid to or allowed by the Contractor.

·3 There shall be added to the net amount paid to or allowed by the Contractor under paragraph B·2·2, in respect of each person employed by the Contractor who is engaged upon or in connection with the Works either on or adjacent to the site and who is not within the definition of workpeople in paragraph B·12·3, the same amount as is payable or allowable in respect of a craftsman under paragraph B·2·2 or such proportion of that amount as reflects the time (measured in whole working days) that each such person is so employed. The provisions of paragraph B·1·4 shall apply to this paragraph B·2·3.

·4 The Contract Sum is based upon the types and rates of refund of the contributions, levies and taxes payable by a person in his capacity as an employer and upon the types and rates of premium receivable by a person in his capacity as an employer being in each case types and rates which at the Base Date are receivable by the Contractor. Such a type and such a rate are in paragraph B·2·5 referred to as a 'tender type' and a 'tender rate'.

·5 If any of the tender rates is increased or decreased or if a tender type ceases to be payable or if a new type of refund of any contribution, levy or tax payable by a person in his capacity as an employer becomes receivable or if a new type of premium receivable by a person in his capacity as an employer becomes receivable after the Base Date, then in any such case the net amount of the difference between what the Contractor actually receives or will receive in respect of workpeople as referred to in paragraphs B·1·1·1 and B·1·1·2 or because of his employment of such workpeople and what he would have received had the alteration, cessation or new type of refund or premium not become effective shall, as the case may be, be paid to or allowed by the Contractor.

·6 The references in paragraphs B·2·4 and B·2·5 to premiums shall be construed as meaning all payments howsoever they are described which are made under or by virtue of an Act of Parliament to a person in his capacity as an employer and which affect the cost to an employer of having persons in his employment.

·7 Where employer's contributions are payable by the Contractor in respect of workpeople as referred to in paragraphs B·1·1·1 and B·1·1·2 whose employment is contracted-out employment within the meaning of the Pension Schemes Act 1993, the Contractor shall, subject to the proviso hereto, for the purpose of recovery or allowance under paragraph B·2 be deemed to pay employer's contributions as if that employment were not contracted-out employment; provided that this paragraph B·2·7 shall not apply where the occupational pension scheme, by reference to membership of which the employment of workpeople is contracted-out employment, is established by the rules of the Construction Industry Joint Council or of some other wage-fixing body so that contributions to such occupational pension scheme are within the payment and allowance provisions of paragraph B·1.

·8 The references in paragraphs B·2·1 to B·2·5 and B·2·7 to contributions, levies and taxes shall be construed as meaning all impositions payable by a person in his capacity as an employer howsoever they are described and whoever the recipient which are imposed under or by virtue of an Act of Parliament and which affect the cost to an employer of having persons in his employment.

Deemed calculation of Contract Sum – materials, goods, electricity and fuels

B·3 The Contract Sum shall be deemed to have been calculated in the manner set out below and shall be subject to adjustment in the events specified hereunder.

·1 The Contract Sum is based upon the market prices which were current at the Base Date of the materials, goods, electricity, fuels or any other solid, liquid or gas necessary for the execution of the Works, and upon the duty or tax payable at that date on the disposal of waste from the site.

·2 If after the Base Date the market price of any of the above things increases or decreases, or the duty or tax on the disposal of waste from the site increases or decreases, then the net amount of the difference shall, as the case may be, be paid to or allowed by the Contractor.

·3 The references in paragraphs B·3·1 and B·3·2 to market price(s) shall be construed as including any duty or tax (other than any VAT which is treated, or is capable of being treated, as input tax by the Contractor) by whomsoever payable which is payable under or by virtue of any Act of Parliament on the import, purchase, sale, appropriation, processing, use or disposal of any of the things described in paragraph B·3·1.

Sub-let work – incorporation of provisions to like effect

B·4 ·1 If the Contractor sub-lets any portion of the Works to a sub-contractor he shall incorporate in the sub-contract provisions to the like effect as the provisions of Fluctuations Option B (excluding this paragraph B·4) including the percentage stated in the Contract Particulars pursuant to paragraph B·13 which are applicable for the purposes of this Contract.

·2 If the price payable under such a sub-contract as referred to in paragraph B·4·1 is increased above or decreased below the price in such sub-contract by reason of the operation of the said incorporated provisions, then the net amount of such increase or decrease shall, as the case may be, be paid to or allowed by the Contractor under this Contract.

Written notice by Contractor

B·5 ·1 The Contractor shall give a written notice to the Architect/Contract Administrator of the occurrence of any of the events referred to in such of the following provisions as are applicable for the purposes of this Contract:

·1 paragraph B·1·2;

·2 paragraph B·1·6;

·3 paragraph B·2·2;

·4 paragraph B·2·5;

·5 paragraph B·3·2;

·6 paragraph B·4·2.

·2 Any notice required to be given under paragraph B·5·1 shall be given within a reasonable time after the occurrence of the event to which the notice relates, and the giving of a written notice in that time shall be a condition precedent to any payment being made to the Contractor in respect of the event in question.

Agreement – Quantity Surveyor and Contractor

B·6 The Quantity Surveyor and the Contractor may agree what shall be deemed for all the purposes of this Contract to be the net amount payable to or allowable by the Contractor in respect of the occurrence of any event such as is referred to in any of the provisions listed in paragraph B·5·1.

Fluctuations added to or deducted from Contract Sum

B·7 Any amount which from time to time becomes payable to or allowable by the Contractor by virtue of paragraphs B·1 to B·3 or paragraph B·4 shall, as the case may be, be added to or deducted from:

·1 the Contract Sum; and

·2 any amounts payable to the Contractor and which are calculated in accordance with clause 8·12·3·1.

The addition or deduction to which this paragraph B·7 refers shall be subject to the provisions of paragraphs B·8 to B·10·1.

Evidence and computations by Contractor

B·8 As soon as is reasonably practicable the Contractor shall provide such evidence and computations as the Architect/Contract Administrator or the Quantity Surveyor may reasonably require to enable the amount payable to or allowable by the Contractor by virtue of paragraphs B·1 to B·3 or paragraph B·4 to be ascertained; and in the case of amounts payable to or allowable by the Contractor under paragraph B·1·3 (or paragraph B·4 for amounts payable to or allowable under the provisions in the sub-contract to the like effect as paragraphs B·1·3 and B·1·4) – employees other than workpeople – such evidence shall include a certificate signed by or on behalf of the Contractor each week certifying the validity of the evidence reasonably required to ascertain such amounts.

No alteration to Contractor's profit

B·9 No addition to or deduction from the Contract Sum made by virtue of paragraph B·7 shall alter in any way the amount of profit of the Contractor included in that Sum.

Position where Contractor in default over completion

B·10 ·1 Subject to the provisions of paragraph B·10·2 no amount shall be added or deducted in the computation of the amount stated as due in an Interim Certificate or in the Final Certificate in respect of amounts otherwise payable to or allowable by the Contractor by virtue of paragraphs B·1 to B·3 or paragraph

continued B·10·1

B·4 if the event (as referred to in the provisions listed in paragraph B·5·1) in respect of which the payment or allowance would be made occurs after the Completion Date.

·2   Paragraph B·10·1 shall not be applied unless:

·1   the printed text of clauses 2·26 to 2·29 is unamended and forms part of the Conditions; and

·2   the Architect/Contract Administrator has, in respect of every written notification by the Contractor under clause 2·28, fixed or confirmed in writing such Completion Date as he considers to be in accordance with that clause.

Work etc. to which paragraphs B·1 to B·4 not applicable

B·11   Paragraphs B·1 to B·4 shall not apply in respect of:

·1   work for which the Contractor is allowed daywork rates under clause 5·7;

·2   changes in the rate of VAT charged on the supply of goods or services by the Contractor to the Employer under this Contract.

Definitions for use with Fluctuations Option B

B·12   In Fluctuations Option B:

·1   the Base Date means the date stated as such in the Contract Particulars;

·2   "materials" and "goods" include timber used in formwork but do not include other consumable stores, plant and machinery;

·3   "workpeople" means persons whose rates of wages and other emoluments (including holiday credits) are governed by the rules or decisions or agreements of the Construction Industry Joint Council or some other wage-fixing body for trades associated with the building industry;

·4   "wage-fixing body" means a body which lays down recognised terms and conditions of workers;

·5   "recognised terms and conditions" means terms and conditions of workers in comparable employment in the trade or industry, or section of trade or industry, in which the employer in question is engaged which have been settled by an agreement or award to which the parties are employers' associations and independent trade unions which represent (generally, or in the district in question, as the case may be) a substantial proportion of the employers and of the workers in the trade, industry or section being workers of the description to which the agreement or award relates.

Percentage addition to fluctuation payments or allowances

B·13   There shall be added to the amount paid to or allowed by the Contractor under:

·1   paragraph B·1·2,

·2   paragraph B·1·3,

·3   paragraph B·1·6,

·4   paragraph B·2·2,

·5   paragraph B·2·5,

·6   paragraph B·3·2

the percentage stated in the Contract Particulars.

Fluctuations Option C

Formula adjustment

Adjustment of Contract Sum – Formula Rules

C·1 ·1 ·1 The Contract Sum shall be adjusted in accordance with the provisions of Fluctuations Option C and the Formula Rules current at the Base Date issued for use with Fluctuations Option C by the JCT ('the Formula Rules').

·2 Any adjustment under Fluctuations Option C shall be to sums exclusive of VAT and nothing in Fluctuations Option C shall affect in any way the operation of clause 4·6.

·2 The Definitions in rule 3 of the Formula Rules shall apply to Fluctuations Option C.

·3 The adjustment referred to in Fluctuations Option C shall be effected (after taking into account any Non-Adjustable Element[70]) in all certificates for payment issued under the provisions of the Conditions.

·4 If any correction of amounts of adjustment under Fluctuations Option C included in previous certificates is required following any operation of rule 5 of the Formula Rules, such correction shall be given effect in the next certificate for payment to be issued.

Interim valuations

C·2 Interim valuations shall be made before the issue of each Interim Certificate.

Fluctuations – articles manufactured outside the United Kingdom

C·3 For any article to which rule 4(ii) of the Formula Rules applies the Contractor shall insert in a list attached to the Contract Documents the market price of the article in sterling (that is the price delivered to the site) current at the Base Date. If after that date the market price of the article inserted in that list increases or decreases then the net amount of the difference between the cost of purchasing at the market price inserted in such list and the market price payable by the Contractor and current when the article is bought shall, as the case may be, be paid to or allowed by the Contractor. The reference to market price in this paragraph C·3 shall be construed as including any duty or tax (other than any VAT which is treated, or is capable of being treated, as input tax by the Contractor) by whomsoever payable under or by virtue of any Act of Parliament on the import, purchase, sale, appropriation or use of the article specified as aforesaid.

Power to agree – Quantity Surveyor and Contractor

C·4 The Quantity Surveyor and the Contractor may agree any alteration to the methods and procedures for ascertaining the amount of formula adjustment to be made under Fluctuations Option C, and the amounts ascertained after the operation of such agreement shall be deemed for all the purposes of this Contract to be the amount of formula adjustment payable to or allowable by the Contractor in respect of the provisions of Fluctuations Option C. Provided always:

·1 that no alteration to the methods and procedures shall be agreed unless it is reasonably expected that the amount of formula adjustment so ascertained will be the same or approximately the same as that ascertained in accordance with Part I or Part II of section 2 of the Formula Rules whichever Part is stated to be applicable in the Contract Documents; and

·2 that any agreement under paragraph C·4 shall not have any effect on the determination of any adjustment payable by the Contractor to any sub-contractor.

Position where Monthly Bulletins are delayed, etc.

C·5 ·1 If at any time prior to the issue of the Final Certificate formula adjustment is not possible because of delay in, or cessation of, the publication of the Monthly Bulletins, adjustment of the Contract Sum shall be made in each Interim Certificate during such period of delay on a fair and reasonable basis.

---

[70] Applies to Local Authorities only.

continued C·5

·2   If publication of the Monthly Bulletins is recommenced at any time prior to the issue of the Final Certificate, the provisions of Fluctuations Option C and the Formula Rules shall apply for each Valuation Period as if no delay or cessation had occurred and the adjustment under Fluctuations Option C and the Formula Rules shall be substituted for any adjustment under paragraph C·5·1.

·3   During any such period of delay or cessation the Contractor and the Employer shall operate such parts of Fluctuations Option C and the Formula Rules as will enable the amount of formula adjustment due to be readily calculated upon recommencement of publication of the Monthly Bulletins.

Formula adjustment – failure to complete

C·6   ·1   ·1   If the Contractor fails to complete the Works by the Completion Date, formula adjustment of the Contract Sum under Fluctuations Option C shall be effected in all Interim Certificates issued after the Completion Date by reference to the Index Numbers applicable to the Valuation Period in which the Completion Date falls.

·2   If for any reason the adjustment included in the amount certified in any Interim Certificate which is or has been issued after the Completion Date is not in accordance with paragraph C·6·1·1, such adjustment shall be corrected to comply with that paragraph.

·2   Paragraph C·6·1 shall not be applied unless:

·1   the printed text of clauses 2·26 to 2·29 is unamended and forms part of the Conditions; and

·2   the Architect/Contract Administrator has, in respect of every written notification by the Contractor under clause 2·28, fixed or confirmed in writing such Completion Date as he considers to be in accordance with that clause.